The Illustrated Dictionary of Nonlinear Dynamics and Chaos

# The Illustrated Dictionary of Nonlinear Dynamics and Chaos

Tomasz Kapitaniak
Technical University of Lodz

Steven R. Bishop
University College London

JOHN WILEY & SONS
Chichester · New York · Weinheim · Brisbane · Singapore · Toronto

Baffins Lane, Chichester
West Sussex PO19 1UD, England

*International* (+44) 1243 779777
*National* (01243) 779777
e-mail (for orders and customer service enquiries): cs-books@wiley.co.uk
Visit our Home Page on http://www.wiley.co.uk
or
http://www.wiley.com



*Other Wiley Editorial Offices*

John Wiley & Sons, Inc., 605 Third Avenue,
New York, NY 10158-0012, USA

WILEY-VCH GmbH, Pappelallee 3,
D-69469 Weinheim, Germany

Jacaranda Wiley Ltd, 33 Park Road, Milton,
Queensland 4064, Australia

John Wiley & Sons (Asia) Pte Ltd, 2 Clementi Loop #02-01,
Jin Xing Distripark, Singapore 129809

John Wiley & Sons (Canada) Ltd, 22 Worcester Road,
Rexdale, Ontario M9W 1L1, Canada

***Library of Congress Cataloging-in-Publication Data***

Kapitaniak, Tomasz.
The illustrated dictionary of nonlinear dynamics and
chaos / Tomasz Kapitaniak & Steven R. Bishop.
p. cm.
Includes bibliographical references.
ISBN 0-471-98323-3 (hardback : alk. paper)
1. Dynamics — Dictionaries. 2. Chaotic behavior in –
Systems – Dictionaries. I. Bishop, Steven R. II. Title.
QC133.K36 1998
531.′11 — dc21 98-29157
CIP

***British Library Cataloguing in Publication Data***

A catalogue record for this book is available from the British Library

ISBN 0-471-98323-3

Produced from Camera Ready Copy supplied by the author.
Printed and bound in Great Britain by Biddles Ltd, Guildford, Surrey.
This book is printed on acid-free paper responsibly manufactured from sustainable forestry, in which at le trees are planted for each one used for paper production.

# Contents

# Preface

The study of dynamical systems actually dates back many years but the last three decades have seen intensive studies which have been prompted by the discovery of chaos. Initially, chaos was seen purely as a mathematical curiosity and elsewhere, although irregular or unpredictable behaviour may have been noted, this was often attributed to random external influences. Correspondingly in engineering, and in particular in electrical systems, the appearance of chaos was usually regarded as a nuisance and thus designed out where possible. Changes came about with publication of seminal works by Lorenz, Feigenbaum, Smale, and May, coupled with numerical simulations by a host of researchers, notably early work by Ueda in electrical engineering, so that modern studies have now confirmed that chaotic phenomena are completely deterministic, occurring in a variety of nonlinear problems in physical and natural systems. Mathematically, the study of chaotic systems has proved extremely useful since the latter form archetypal dynamical systems exhibiting various types of interesting behaviour, some of which remains unexplained even today. The intriguing properties and tantalising possibilities of chaos have thus created considerable interest in the mathematics world, thus leading to a mass of new definitions and results in the general field of nonlinear dynamics which encompasses chaos and systems theory.

Outside of the field of mathematics the natural question to ask is how can we utilise the properties of chaos to our practical advantage? Answering this question forms one of the most active fields of research today, being carried out by scientists in a wide variety of subject areas. On the one hand, chaotic motion is unpredictable and 'random-like' opening up possibilities e.g. for encouraging the creation of an effective tool in ergodic, mixing properties, in improving chemical reactions. Meanwhile, embedded within a chaotic system lies a myriad of periodic solutions whose behaviour may prove to be useful in certain circumstances. Implementation of specifically adapted techniques to control the system from the irregular chaotic motion on to one or more of these more desirable solutions is proving beneficial in a number of physical experiments, thus causing growing interest among engineers and applied scientists. The fact that a large number of potentially useful solutions are available for control means that the same system set-up can be used for a selection of outcomes, so providing greater flexibility.

As a consequence of this dual front of research, advances have been made in various disciplines. The cross fertilisation of ideas, while wholly commendable, means that names and methods are used and developed which may be altered when 'rediscovered' in a different context. Correspondingly, many of the terms now in place stem from different sources so that one drawback encountered by researchers and students in this rapidly expanding field is the proliferation of new terminology.

There is now a staggering array of elementary, and not so elementary, texts available which cover various aspects of dynamics, with each giving some background, while specifically covering in more detail the research areas of interest to the writers. Researchers and students who are new to the field therefore have difficulty in finding one single text which explains all of the concepts that they require.

Our aim here is to help students and researchers coming from a broad scientific base by listing many (but obviously not all) of the equations, terms, theorems, etc., which arise in the study of nonlinear dynamical systems. In so doing, new mathematical ideas are described and explained with examples, including illustrative figures where they aid the understanding. In many cases, remarks are added to underline important aspects and simplify the meaning where possible. On the other hand, the need for mathematical rigour is emphasised with precise definitions. For some readers, these brief descriptions will be sufficient to allow them to continue their own investigations but, should more detail or proofs be required, further references are given.

In compiling this dictionary, where alternative terms are possible an entry is placed under the name in most common usage, with cross-references being given under other names. In addition, if it is considered that further understanding will accrue, then full additional cross-referenced entries are provided.

This dictionary is generally appropriate for postgraduate students and researchers in engineering, mathematics, physics and applied sciences who have some introductory background in the theory of vibrations and dynamical systems. For undergraduates, some mathematical knowledge would be required for particular entries; specifically, elementary courses in ordinary differential equations and linear algebra. The overall approach is to apply mathematical rigour in an engineering context, naturally reflecting the authors styles — an approach which we hope readers will find useful.

In developing this dictionary the authors have greatly benefited from various discussions with a number of colleagues. Tomasz Kapitaniak would like to thank John Brindley, Leon O. Chua, Celso Grebogi, Mohamed S. El Naschie, Tom Mullin, Maciej Ogorzalek, Willi-Hans Steeb and Jerzy Wojewoda for helpful suggestions. Meanwhile, Steve Bishop would like to note his appreciation for comments made by colleagues at UCL, particularly

Jaroslav Stark, Rua Murray, Gert van der Heijden, Riccardo Carretero Gonzales and Giovanni Santoboni. This said, the final responsibility for interpretation lies entirely with ourselves.

Tomasz Kapitaniak and Steven Bishop
July 1998.

# List of Symbols

| | |
|---|---|
| $A^T$ | transpose of the matrix $A$ |
| $\det A$ | determinant of the matrix $A$ |
| $\mathcal{C}$ | the set of complex numbers |
| $\mathcal{C}^n$ | the $n$-dimensional complex linear space |
| $C^r$ | class of $r$ differentiable functions |
| $d_c$ | capacity dimension |
| $d_{corr}$ | correlation dimension |
| $d_H$ | Hausdorff dimension |
| $d_i$ | information dimension |
| $d_L$ | Lyapunov dimension |
| $d_T$ | topological dimension |
| $f^{(n)}$ | $n$th iteration of the map $f$ |
| $f^{-1}$ | inverse of the map $f$ |
| $f \circ g$ | composition of mappings $(f \circ g)(x) = f(g(x))$ |
| $< f(t) >$ | average value of $f(t)$ |
| $i$ | $\sqrt{-1}$ |
| $I$ | identity matrix |
| $\mathrm{Im}\ x$ | imaginary part of the complex number $x$ |
| $\lambda$ | Lyapunov exponent or eigenvalue |
| $L^p$ | $L^p$ space (family of all integrable functions) |
| $L^1$ | the space of functions $f$ with $\|f\|$ integrable in the sense of Lebesque |
| $\mu(S)$ | Lebesque measure of the set $S$ |
| $m(A)$ | probability measure of the set $A$ |
| $\mathcal{Q}$ | the set of rational numbers |
| $\mathcal{R}$ | the set of real numbers |
| $\mathcal{R}^+$ | non-negative real numbers |
| $\mathcal{R}^n$ | the $n$-dimensional real linear space |
| $\mathrm{Re}\ x$ | real part of the complex number $x$ |
| $S_1$ | circle of length T |
| $\mathrm{sgn}(x)$ | sign function (1 if $x > 0$ and -1 if $x < 0$) |
| $t$ | time |
| $T$ | period of function |

| | |
|---|---|
| $T_x M$ | the tangent space of the manifold $M$ at the point $x$ |
| $U(x)_{[a,b]}(t)$ | unit step function<br>$U(x) = 1$ for $x \in [a,b]$ and $U(x) = 0$ for $x \notin [a,b]$ |
| $W^s$ | stable manifold |
| $W^u$ | unstable manifold |
| $\mathcal{Z}$ | the set of integers |
| $\|\|.\|\|$ | norm |
| $\wedge$ | exterior product (let $a, b \in \mathcal{R}^2$ then $a \wedge b = a_1 b_2 - a_2 b_1$) |

# A

**$\alpha$-limit set** (*see* **alpha-limit set**)
The point $p$ belongs to an $\alpha$-limit set of an orbit $\gamma$: $\alpha(\gamma)$ if $p$ is an $\alpha$-limit point (*see* **limit point**).

**$\alpha$-pseudo orbit**
Consider a one dimensional map

$$x_{n+1} = f(x_n, c)$$

where $x \in \mathcal{R}$, and $c \in \mathcal{R}$ is a control parameter. An orbit is given by the sequence of iterates $x \to f(x, c) \to f^{(2)}(x, c) \to \ldots$, etc. A change of a single digit in $x$ can yield an entirely different orbit which diverges exponentially from that generated by the original $x$.

Let $[0, 1]$ be the phase space and $x$ be a digit string, i.e.

$$x = \sum_{i=1}^{\infty} \frac{\epsilon_i}{\mu^i} = \epsilon_1 \epsilon_2 \ldots \epsilon_N \ldots$$

where $\epsilon_i = 0, 1, 2, \ldots, \mu - 1$ and $\mu \geq 2$ is an integer. Then a positive Lyapunov exponent $\lambda$ (*see* **Lyapunov exponents**) can be understood as

$$\lambda N \sim \ln \frac{\mu}{n}$$

where $n$ is the number of iterations that may be performed before an error in the $N$-th digit of $x$ creates an error in the leading digit of the iterate $f^{(n)}(x, c)$.

By a computable $\alpha$-pseudo orbit we mean any computable sequence $x_n$, such that

$$|f(x_n, c) - x_{n+1}| < \alpha$$

with $n \geq 0$ and the difference

$$f(x_n, c) - x_{n+1} = \delta(x_n, c)$$

being due to a deterministic truncation or round-off error.

The map $f(x_n, c)$ is said to have the shadowing property if for every $\beta > 0$ there is an $\alpha(\beta)$ such that every $\alpha$-pseudo orbit $x_n$ can be $\beta$-shadowed by some exact orbit $\bar{x}_n$, as defined by

$$|\bar{x}_n - x_n| \leq \beta$$

for all $n \geq 0$. If this holds, then $\bar{x}_n$ is said to be a $\beta$-shadow of the pseudo orbit $x_n$.

*Reference*: McCauley and Palmore (1986).

**Abel equation**
The first order ordinary differential equation

$$\frac{\mathrm{d}x}{\mathrm{d}t} = a_0(t) + a_1(t)x^2 + a_2(t)x^3$$

where $x \in \mathcal{R}$ and $a_{0,1,2}(t)$ are continuous bounded functions of time $t$, is called the Abel equation.

**abrupt bifurcation**
A bifurcation at the transition from regular to chaotic scattering (*see also* **bifurcation, chaotic scattering**).

**abstract dynamics**
The concept of viewing dynamics in the abstract form of pure mathematics.

*Example*: A dynamical system can be defined in the following way. Let $M$ be a smooth manifold, $\mu$ a measure on $M$ defined by a continuous positive density, and $f^t : M \rightarrow M$ a one-parameter group of measure-preserving diffeomorphisms. The collection $(M, \mu, f^t)$ is then called a classical dynamical system.

*Reference*: Arnold and Avez (1968).

**action-angle variables**
Consider a conservative Hamiltonian system with $N$ degrees of freedom so that the phase space is $\mathcal{R}^{2N}$ with canonical co-ordinates $p_i, q_i, where i = 1, 2, ..., N$. Let $H$ be the Hamiltonian function and assume that the system evolution has $N$ first integrals (*see* **first integral**), $F_1 = H$, $F_2, ..., F_N$ which are in involution, i.e. for all $i, j$

$$\frac{\partial F_i}{\partial q_i}\frac{\partial F_j}{\partial p_j} - \frac{\partial F_i}{\partial p_i}\frac{\partial F_j}{\partial q_j} = 0.$$

The constants of motion are given by the hypersurfaces, $F_1 = f_1, ..., F_N = f_N$. The real constants $f_1, ..., f_N$ are given by the initial values $p(0), q(0)$. The construction of action-angle variables in $\mathcal{R}^{2N} = \{(p, q)\}$ is as follows:

Let $\gamma_1, ..., \gamma_N$ be a basis for the one-dimensional cycles on the torus $M_f$ (the increase of the coordinate $\phi_i$ on the cycle $\gamma_j$ is equal to $2\pi$ if $i = j$ and 0 if $i \neq j$). The coordinate $\phi$ $(= (\phi_1, ...\phi_N))$ on $M_f$ is called an angle variable. If we now define

$$I_i(f) = \frac{1}{2\pi}\oint_{\gamma_i} \mathrm{p}\mathrm{d}q = \frac{1}{2\pi}\int_{\gamma_i} (p_i \mathrm{d}q_i + ... + p_N \mathrm{d}q_N)$$

then the $N$ quantities $I_i(f)$ so defined are called the action variables.

*Remarks*:

(i) The transformation $(p, q) \to (I, \phi)$ is canonical, i.e.

$$\sum_{j=1}^{N} dp_j \wedge dq_j = \sum_{j=1}^{N} dI_j \wedge d\phi_j$$

where $\wedge$ denotes the exterior product $(dq_i \wedge dq_j = -dq_j \wedge dq_i)$.

(ii) For the given values $f_i$ of the $N$ first integrals $F_i$, if the $N$ quantities $I_i$ are independent, then in the neighbourhood of the torus $M_f$ we can take the variables $(I, \phi)$ as coordinates.

(iii) The action-angle variables are not uniquely defined.

*Example*: Let $N = 1$. We can take

$$I' = I + \text{constant}$$

for the action variable and

$$\phi' = \phi + c(I)$$

for the angle variable.

*References*: Arnold (1988); Steeb (1991).

**adaptive control**
A term used in control theory. A control procedure which adapts to slow changes in the system and which is not sensitive to the external perturbations is called adaptive.

*Reference*: van Campen (1997).

**adiabatic invariant**

Quantities which are asymptotically preserved under a sufficiently slow variation of the parameters of a Hamiltonian system are called adiabatic invariants.

*Example*: Consider the Hamiltonian system

$$\frac{du}{dt} = V(u, \lambda)$$

where $\lambda \in \mathcal{R}$ is a parameter. A function $I$ of the pair consisting of the point $u$ in the phase space and the parameter $\lambda$, is called an adiabatic invariant if, for any smooth function $\lambda(\tau)$ of the slow time $\tau = \epsilon t$, the variation of $I(u(t), \lambda(\epsilon t))$ along a solution of the equation

$$\frac{du}{dt} = V(u, \lambda(\epsilon))$$

remains small within the time interval $0 \leq \tau \leq 1/\epsilon$ provided that $\epsilon$ is sufficiently small.

*References*: Arnold (1988), Steeb (1991).

**affine map**

A map $f : \mathcal{R}^2 \to \mathcal{R}^2$ which is of the form $f(x_1, x_2) = (ax_1 + bx_2 + e, cx_1 + dx_2 + f)$, where $a$–$f$ are real numbers, is called a two-dimensional affine map.

An affine map can be generalised to any dimension.

*Reference*: Falconer (1985).

**algorithmic complexity**

A sequence, $x(n) = x_0, ..., x_{n-1}$, is defined to be random:

(i) if it passes statistical tests of randomness;

or

(ii) if there exists no finite size description of a 'law' which is able to reproduce the sequence with arbitrary length.

The latter requirement can be represented in terms of algorithmic complexity theory. The algorithmic complexity $\mathcal{H}(x(n))$ of a sequence $x(n)$ is the minimal program length necessary to output $x(n)$ on a computer, i.e. if $p$ symbolises a program running on a computer model $C$, then

$$\mathcal{H}(x(n)) = \min_{C(p)=x(n)} \text{length}(p).$$

A sequence is defined to be random if, as $x(n)$ increases in length $n$, $\mathcal{H}(x(n))$ increases in such a way that

$$\lim_{n\to\infty} [n - H(x(n))] < \infty.$$

*Remark*: The above definition implies that a random sequence cannot be substantially 'compressed' by computational effort.

*References*: Chaitin (1987); Steeb (1991).

**almost-periodic function**
A vector or matrix function $f(t)$, defined and continuous for $-\infty < t < \infty$, is said to be almost periodic in $t$ if for any $\epsilon > 0$ there exists an $l = l(\epsilon) > 0$ such that in any interval of length $l$ there is the so-called translation number $\tau$ independent of $t$, such that

$$|f(t+\tau) - f(t)| < \epsilon$$

for all $t \in \mathcal{R}$.

A function $f(x,t)$, defined and continuous in $\Lambda \times \mathcal{R}$, where $\Lambda$ is a compact subset of $\mathcal{R}^n$, is called almost periodic in $t$ uniformly with respect to $x$ in $\Lambda$, if the quantities $l$ and $\tau$ in the definition of an almost-periodic function can be chosen independently of $x \in \Lambda$.

*Reference*: Kapitaniak (1991).

**Andronov–Hopf bifurcation** (*see* **Hopf bifurcation**)

**Anosov diffeomorphism**
Let $f : M \to M$ be a diffeomorphism of a compact manifold $M$. Assume that:

(i) The tangent space of $M$ is decomposed into the direct sum of two subspaces at every point of $M$, i.e.

$$T_x M = X_x \oplus Y_x$$

where $x \in M$ and $\oplus$ denotes the direct sum.

(ii) The fields of the planes $X = \{X_x\}$ and $Y = \{Y_x\}$ are continuous and invariant with respect to $A$.

(iii) For some Riemannian metric, the map $A$ contracts the planes of the first field and expands the planes of the second field. This means that there exists a number $\lambda < 1$ such that for any point $x$ of $M$

$$|A_*\zeta|| \leq \lambda||\zeta|| \text{ and } ||A_*\eta|| \leq \lambda^{-1}||\eta||$$

for all $\zeta \in X_x$ and $\eta \in Y_x$. Then $f$ is called an Anosov diffeomorphism.

*Reference*: Arnold (1988).

**Anosov system**

Let $M$ be a compact smooth manifold, $V$ a vector field on $M$ without fixed points and $\{g^t\}$ the corresponding phase flow. Assume that:

(i) At each point of $M$, the tangent space of $M$ can be represented as the direct sum of three subspaces

$$T_x M = X_x \oplus Y_x \oplus Z_x$$

for $x \in M$.

(ii) The vector fields $X, Y$ and $Z$ of the planes are continuous and invariant with respect to the phase flow $\{g^t\}$.

(iii) The vector field $Z$ is generated by the field of phase velocity.

(iv) For some positive constants, $c$ and $\lambda$, and for some Riemannian metric on $M$, we have

$$||g^t|_X|| \leq ce^{-\lambda t}$$

for $t > 0$, and

$$||g^t|_Y|| \leq ce^{\lambda t}.$$

Then the phase flow is said to be an Anosov flow and the ordinary differential equation

$$\frac{\mathrm{d}x}{\mathrm{d}t} = V(x)$$

an Anosov system.

*Reference*: Arnold (1983).

**Anosov theorem**

This theorem says that any Anosov system is structurally stable.

*Example*: The torus automorphism $A : T^2 \to T^2$ defined by the matrix

$$\begin{pmatrix} 2 & 1 \\ 1 & 1 \end{pmatrix}$$

is structurally stable in the $C^1$-topology, i.e. every diffeomorphism $B$ which is sufficiently close to $A$ and the derivative of $B$ sufficiently close to the derivative of $A$, is conjugate with $A$ by means of some homeomorphism $H$ so $B = H^{-1}AH$.

*Reference*: Arnold (1983).

**aperiodic function**
A function $f$ which is not periodic is said to be aperiodic.

**Appolonian gasket**
The Appolonian gasket is a subset of the plane $\mathcal{R}^2$ obtained as a limit of the sets $C_n$: $n = 0, 1, \ldots$, in the following construction. As the first step, three mutually tangent circles with radius 1 are plotted. The set $C_0$ is the area enclosed by three arcs (see Figure 1). Each $C_n$ consists of some regions bounded by three mutually tangent circular arcs. To obtain $C_{n+1}$, remove from $C_n$ a circle which is tangent to all three of the arcs. The boundary of the circle remains so the Appolonian gasket is the limit intersection of the sets $C_n$.

*Reference*: Steeb (1991).

**approximation methods**
There are several traditional methods for finding approximate solutions to dynamical systems which can be split into either approximation methods (weighted residual, Galerkin, **harmonic balance** and finite element methods) or perturbation methods (expansion, Linstedt-Poincaré, multiple scales, averaging methods).

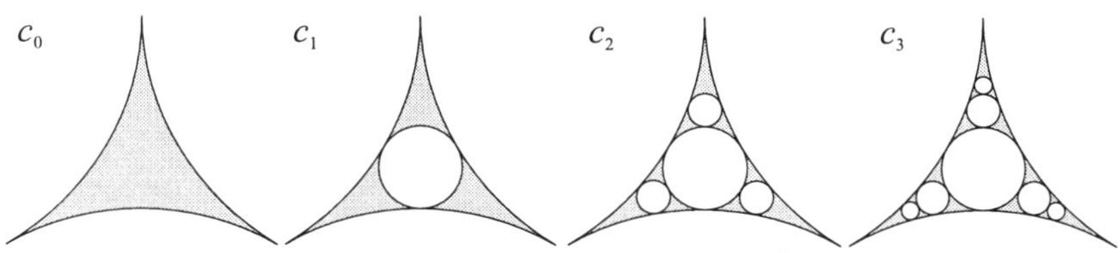


**Figure 1** First steps in the construction of the Appolonian gasket

**area–preserving map** (*see* **volume preserving map**)

**Arnold cat map** (*see* **cat map**)

**Arnold diffusion**
Let $N$ be the number of degrees of freedom of a Hamiltonian system. If $N > 2$ then the $N$-tori are not separatrices in the $(2N-1)$-dimensional surfaces $H(p,q) = E$, so a chaotic orbit is not constrained by invariant tori. In numerical studies, it is found that the evolution of the action variable apparently has no directional character, i.e. it represents a 'random-like' wandering in the resonance region between the invariant tori. This process is called Arnold diffusion.

*Reference*: Arnold (1988).

**Arnold tongues**
Consider the circle map

$$\theta_{n+1} = \theta_n + \Omega - \frac{K}{2\pi} \sin 2\pi\theta_n \bmod 1,$$

where $\Omega$ and $K$ are constant. The rotation number (*see also* **winding number**) is defined as follows:

$$\rho = \lim_{N\to\infty} \frac{1}{N} \sum_{n=1}^{N} |\theta_n - \theta_{n-1}| \bmod 1,$$

when the limits exists. The $(K, \Omega)$ parameter space of this system is characterised by the fact that for $K < 1$ the limit always exists, but can be either a rational or irrational number.

The regions in $(K, \Omega)$ parameter space where unique rational values of $\rho$ exist are called Arnold tongues.

*Remark*: Arnold tongues are characteristics of periodically forced systems.

*Reference*: Arnold (1983).

**artificial neural networks** (*see* **neural networks**)

**asymmetric tent map**
The map $f : [0,1] \to [0,1]$, defined as

$$f(x) = \begin{cases} ax + (a+b-ab)/b & : \quad 0 \le x \le c \\ b(1-x) & : \quad c \le x \le 1 \end{cases}$$

where $c = (b-1)/b$ , $a \geq 0$, $b \geq 1$ and $a + b \geq ab$, is called the asymmetric tent map or skew-tent map.

**asymptotic stability** (*see* **stability**, **attractor**)

**attractor**

Despite its obvious importance, a strict mathematical definition of an attractor is not universally agreed upon. It is possible, however, to write definitions which convey much of the concept.

The invariant subset $A$ of a phase space $\mathcal{R}^n$ of the differential equation

$$\frac{\mathrm{d}u}{\mathrm{d}t} = f(u)$$

where $u \in \mathcal{R}^n$, which is reached asymptotically as $t \to \infty$ ($t \to -\infty$), is called an attractor (repellor).

An attractor $A$ is stable if it is Lyapunov-stable and its basin of attraction $\beta(A)$ has positive Lebesgue measure (*see* **basin of attraction**).

An attractor $A$ is asymptotically stable if it is Lyapunov-stable and $\beta(A)$ contains a neighbourhood of $A$.

*Remark*: The difference between Lyapunov-stable and asymptotically stable attractors is illustrated in Figures 2 and 3.

*Example*: For a three-dimensional dissipative system whose steady state solutions are governed by three eigenvalues: $\lambda$: ($\lambda_1$, $\lambda_2$, $\lambda_3$) there are three possible stable types of solutions, namely $\lambda$: $(-,-,-)$ point-like attractor, $\lambda$: $(0,-,-)$ limit cycle, and $\lambda$:$(0,0,-)$ focus two-frequency quasi-periodicity. In

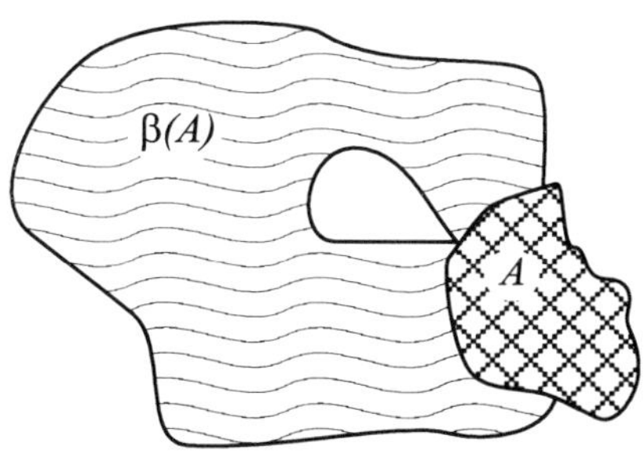


**Figure 2** Lyapunov-stable attractor

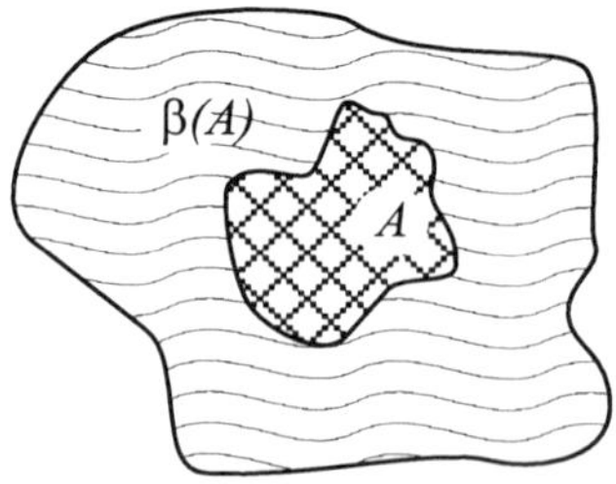


**Figure 3** Asymptotically stable attractor

addition, there are non-trivial attractors with $\lambda$: $(+, 0, -)$ whenever

$$\sum_{i=1}^{n} \lambda_i < 0.$$

Attractors with positive Lyapunov exponents (*see* **Lyapunov exponents**) are called chaotic attractors, and the solution of the differential equation is said to be chaotic if at least one one-dimensional Lyapunov exponent is positive.

*Remark*: In a chaotic attractor, the positive Lyapunov exponent indicates exponential spreading within the attractor in the direction transverse to the flow, while the negative exponent indicates exponential contraction on to the attractor. Under the action of such a flow, phase-space volumes evolve into sheets, as shown in Figure 4.

Exponential divergence of nearby trajectories within a compact subspace requires a folding action. A simple example of this process is shown in Figure 5. Trajectories first diverge exponentially within a sheet, then the sheet folds and then connects back to itself ($A$ is connected with $A^1$, and $B$ with $B^1$), thus

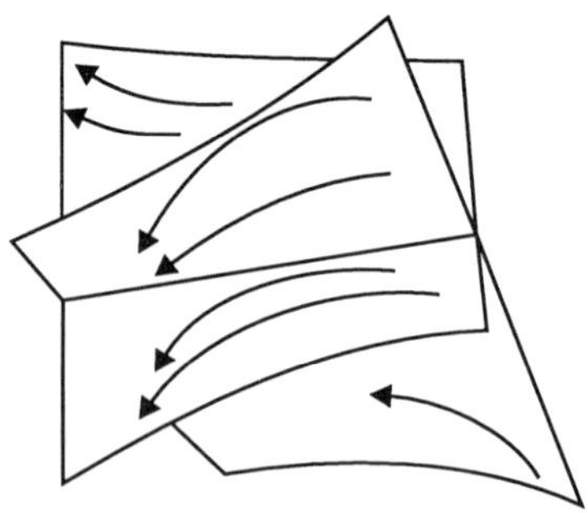

**Figure 4** Exponential spreading and contraction of trajectories on an attractor with $(+, 0, -)$ Lyapunov exponents

producing the attractor shown in Figure 6.

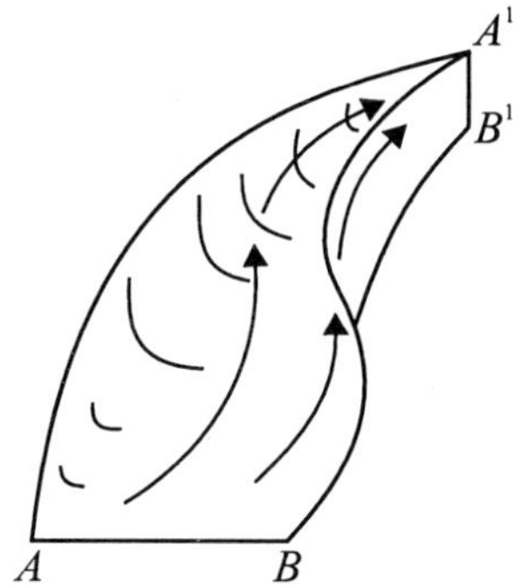


**Figure 5** Folding of sheets

Generally, this type of attractor is not simply a sheet with a single fold, but a sheet folded and refolded infinitely by a flow.

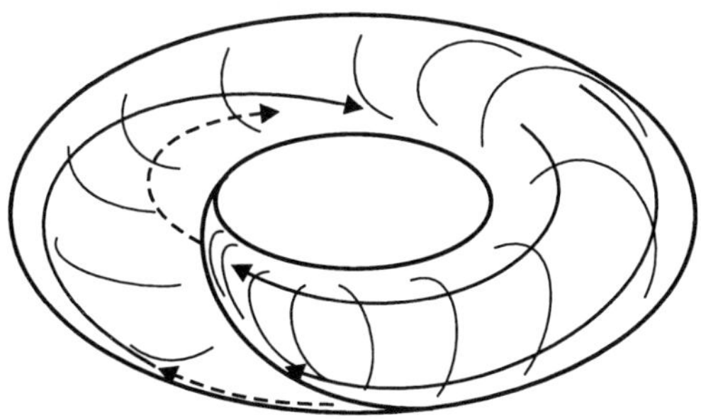

**Figure 6** Attractor obtained after connection of $A \to A^1$ and $B \to B^1$

*References*: Guckenheimer and Holmes (1983); Kapitaniak (1991); Milnor (1985); Ruelle (1981); Thompson and Stewart (1986).

### Aubry–Mather theorem

Let $f : \mathcal{S}^1 \to \mathcal{S}^1$ be a homeomorphism of the circle $\mathcal{S}^1$ and suppose that there is a continuous function $\bar{f} : \mathcal{R} \to \mathcal{R}$ such that $\pi(\bar{f}(x)) = f(\pi(x))$, where $\pi(x) = x \bmod 1 = \theta$ ($\theta$ is the angular displacement at the centre of the circle related to a reference radius $r$). Then $\bar{f}$ is called a lift of $f : \mathcal{S}^1 \to \mathcal{S}^1$ on to $\mathcal{R}$.

Let $f$ be a twist map. A closed set, then the $f$-invariant set $E \subset A$ is a Mather set if the following apply:

(i) $E$ is the graph of a continuous function $\Phi$ defined on a closed subset $K$ of a circle $\mathcal{S}^1$ taking values in [0,1];

(ii) the lift $\bar{f}$ preserves the order on the covering of $E$.

*Theorem*:
Let $f$ be an area-preserving twist homeomorphism of the annulus $A = \mathcal{S}^1 \times [a, b]$. Then, for every $\rho \in [a, b]$, $f$ has a Mather set with rotation number $\rho$ (*see* **winding number**).

*Reference*: Arrowsmith and Place (1990).

**autocatalytic system** (*see* **self-exciting system**)

**autocorrelation function**

*1. Discrete function*:
Let $X$ be an open subset of $\mathcal{R}$. Then the mapping $f : X \to X$ can be written as the difference equation

$$x_{n+1} = f(x_n)$$

where $x_0 \in X$ is the initial value and $n = 0, 1, 2, ...,$ etc. Let

$$\langle x_n \rangle = \lim_{T\to\infty} \frac{1}{T} \sum_{n=0}^{T} x_n$$

be the time average of $x_n$, where $x_n$ and $\langle x_n \rangle$ depend upon $x_0$. The autocorrelation function $C_{xx}$ is defined as

$$C_{xx}(\tau) = \lim_{T\to\infty} \frac{1}{T} \sum_{n=0}^{T} (x_n - \langle x_n \rangle)(x_{n+\tau} - \langle x_n \rangle)$$

where $\tau = 0, 1, 2, ...,$ and $C_{xx}$ depends on the initial value $x_0$.

*2. Continuous function*:
Let $f(t)$ be a bounded function for $t \in \mathcal{R}^+$; the autocorrelation function is then defined as

$$C_{ff}(\tau) = \frac{\lim_{T\to\infty} \frac{1}{T} \int_0^T (f(t) - \langle f(t) \rangle)(f(t+\tau) - \langle f(t) \rangle) \mathrm{d}t}{\lim_{T\to\infty} \frac{1}{T} \int_0^T (f(t) - \langle f(t) \rangle)^2 \mathrm{d}t}$$

where the time average $\langle f(t) \rangle$ is given by

$$\langle f(t) \rangle = \lim_{T\to\infty} \frac{1}{T} \int_0^T f(t) \mathrm{d}t.$$

*Example*: Let $f(t) = \sin \Omega t$, then the autocorrelation function is $C_{ff} = \cos \Omega \tau$.

*Reference*: Eckmann and Ruelle (1985).

**automorphism**
This term is used to describe a map $S$ which is measure-preserving and invertible (*see also* **homeomorphism**, **measure-preserving**).

*Reference*: Lasota and Mackey (1985).

**autonomous system**
Consider the first-order system of ordinary differential equations

$$\frac{\mathrm{d}u}{\mathrm{d}t} = f(u), \qquad u(t_0) = u_0$$

where $t \in \mathcal{R}^+$ and $D$ is an open subset of $\mathcal{R}^n$, ($u \in D \subset \mathcal{R}^n$). A system of differential equations of this form, in which the independent variable $t$ does not occur explicitly, is called autonomous.

*Example*: Physical systems to which no external energy is supplied during the motion (e.g. the simple pendulum and the Lorenz model) or constant energy (self-excited systems) are autonomous.

**averaging methods**
Consider a nonlinear differential equation of the form

$$\frac{\mathrm{d}^2 u}{\mathrm{d}t^2} + u = \epsilon F\left(u, \frac{\mathrm{d}u}{\mathrm{d}t}\right) \tag{1}$$

where $\epsilon$ is a small real parameter $0 < \epsilon \ll 1$. If $\epsilon$ is sufficiently small then one can assume a solution of the form

$$u(t) = a(t)\cos(t + \phi(t)) \tag{2}$$

where $a$ and $\phi$ are slowly varying functions of time compared to the fast variables $u$ and $\mathrm{d}u/\mathrm{d}t$. Differentiating this approximation (2) and substituting into the differential equation (1) leads to a further set of equations which may be solved for $\mathrm{d}a/\mathrm{d}t$ and $\mathrm{d}\phi/\mathrm{d}t$. These may be approximately solved using the first approximation introduced by Krylov and Bogoliubov by expanding the components $F\sin(t + \phi(t))$ in a Fourier series and neglecting all terms except the first.

*Reference*: Huntley and Johnson (1983).

**Axiom-A diffeomorphism**

Let $M$ be a compact smooth manifold, $T_xM$ the tangent space of $M$ at the point $x$, and $f : M \to M$ a diffeomorphism. In addition, let $A$ be the set of all non-wandering points $a \in M$ (*see* **wandering point**). If the non-wandering set $A$ is hyperbolic (*see* **hyperbolic set**) and if the periodic points of the map $f$ are dense in $A$, then $f$ is called an Axiom-A diffeomorphism.

*Reference*: Eckmann and Ruelle (1985).

**Axiom-A flow**

Consider the flow $f^t$ ($t \in \mathcal{R}$) on the $m$-dimensional manifold $M$. Let $T_xM$ be the tangent space of $M$ at the point $x$, and $A$ a set of all non-wandering points $a \in M$ (*see* **wandering point**). If the non-wandering set $A$ is hyperbolic (*see* **hyperbolic set**), and if the periodic orbits and fixed points are dense in $A$, then $f^t$ is called an Axiom-A flow.

*Reference*: Eckmann and Ruelle (1985).

**axon**

An axon is a part of a nerve cell which can display chaos. Physical experiments show that the giant squid axon undergoes a self-sustained oscillation when stimulated by a sinusoidal force. In the rest state, when excited by a pulse train force, the giant squid axon has been also shown to behave chaotically.

*Reference*: Winfree (1988).

# B

**Bäcklund transformation**
Suppose that we have two uncoupled partial differential equations, in two independent variables $x$ and $t$. In addition, for the two functions $u$ and $v$, assume that the two equations are expressed as $P(u) = 0$ and $Q(v) = 0$, where $P$ and $Q$ are two operators which are, in general, nonlinear. Let $R_i = 0$ be a pair of relations

$$R_i(u, v, \frac{\mathrm{d}u}{\mathrm{d}x}, \frac{\mathrm{d}u}{\mathrm{d}t}, \frac{\mathrm{d}v}{\mathrm{d}x}, \frac{\mathrm{d}v}{\mathrm{d}t}, ...; x, t) = 0$$

with $i = 1, 2$, between the two functions $u$ and $v$. Then $R_i$ is a Bäcklund transformation if it is integrable for $v$ when $P(u) = 0$ and if the resulting $v$ is a solution of $Q(v) = 0$, and vice versa.

If $P = Q$, so that $u$ and $v$ satisfy the same relationship, then $R_i = 0$ is called an auto-Bäcklund transformation.

*Reference*: Drazin and Johnson (1989).

**baker map**
The map $f : [0, 1] \times [0, 1] \to [0, 1] \times [0, 1]$, where

$$f(x, y) = \begin{cases} (2x, \frac{1}{2}y) & for \quad 0 \le x \le \frac{1}{2} \\ (2x - 1, \frac{1}{2}(y + 1)) & for \quad \frac{1}{2} < x \le 1 \end{cases}$$

is called the baker map. The dynamics of this two-dimensional map are akin to the actions used when kneading dough in the bread making process. First, the unit square is flattened and stretched, and is then folded.

**basin of attraction**
In dissipative nonlinear systems, it is possible for more than one attractor to exist for a single parameter setting. Different initial conditions will evolve towards one or other of the co-existing attractors. The closure of the set of initial conditions which approaches a given attractor is called the basin of

attraction of that attractor. In the case of two or more co-existing attractors, the boundary between one basin of attraction and another is called the basin boundary (*see* Figure 7).

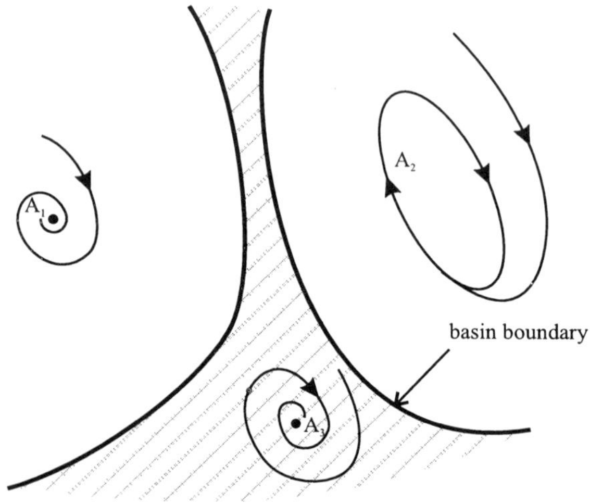


**Figure 7** Schematic representation of basins of attraction of three co-existing attractors, $A_1, A_2$ and $A_3$

*Example 1*: Consider the case of a particle moving in one dimension (along $x$) under the action of friction and the symmetric two-well potential $V(x)$ shown in Figure 8(a). Trajectories from almost every initial condition come to rest at one of the stable fixed points, i.e. $x = x_0$ or $x = -x_0$. In Figure 8(b), we show schematically the basins of attraction for these two attractors in the $(x, \mathrm{d}x/\mathrm{d}t)$ phase plane of the system. Initial conditions starting in the cross-hatched region decay on to the attractor at $x = x_0$, $\mathrm{d}x/\mathrm{d}t = 0$, while initial conditions starting in the uncrossed (blank) region decay on to the attractor at $x = -x_0$, $\mathrm{d}x/\mathrm{d}t = 0$. In this case, the basins of attraction are separated by a simple curve (the basin boundary). This curve passes through the unstable fixed point, $x = \mathrm{d}x/\mathrm{d}t = 0$. The initial conditions on the basin boundary generate a trajectory that approaches the unstable fixed point, i.e. the basin boundary is the stable manifold of an unstable fixed point, which in this case forms the separatrix between the two attractors.

*Example 2*: Consider the resonance curve of Duffing's equation

$$\frac{\mathrm{d}^2 y}{\mathrm{d}t^2} + a\frac{\mathrm{d}y}{\mathrm{d}t} + y - by^3 = \cos(\omega t + \phi) \qquad (1)$$

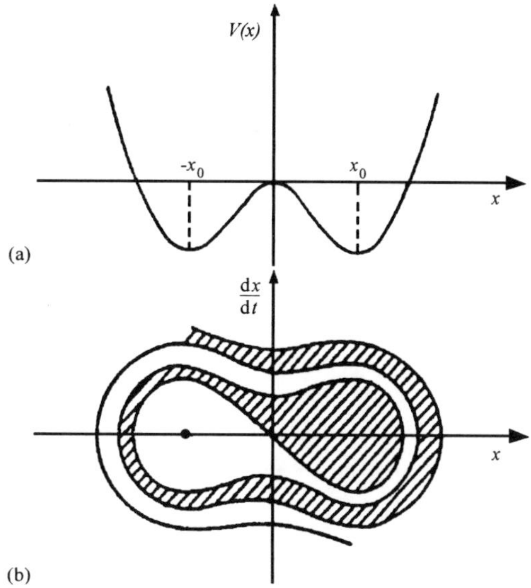


**Figure 8** (a) Potential $V(x)$ for a point particle moving in one dimension. (b) The basins of attraction for the attractors at $x = x_0$ (cross-hatched) and $x = -x_0$ (blank)

with $b > 0$ (shown in Figure 9). Three basic types of final states are possible for this system:

(i) large amplitude oscillations on the upper branch (resonant);

(ii) small amplitude oscillations on the lower branch (non-resonant);

(iii) trajectories which diverge, or escape to infinity.

Basins of attraction of (i)–(iii) attractors are indicated in Figure 10.

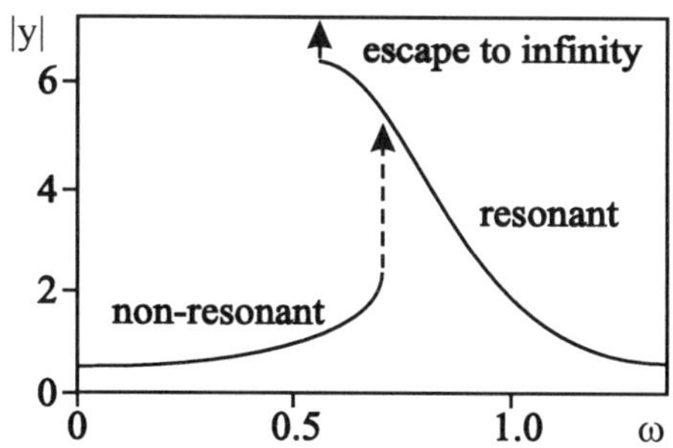


**Figure 9** Resonance curve of Duffing's equation, with $b > 0$

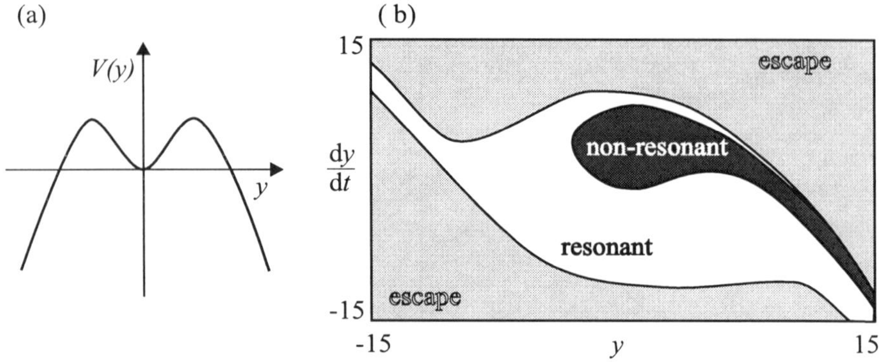


**Figure 10** (a) Potential function and (b) basins of attraction for a Duffing oscillator

In these examples, as well as in Figure 7, the basin boundaries are smooth, continuous lines. This implies that when the initial conditions are away from the boundary, small perturbations will not qualitatively affect the response. However, it has been shown that in nonlinear systems, this boundary may change so that it is not smooth but has a fractal structure; in this case it is called a fractal basin boundary. Of course in this case, any small uncertainties in initial conditions may lead to uncertainties in the outcome of the system, with this being termed as final-state sensitivity.

*Example 3*: Figure 11 shows the basin structure of the forced damped pendulum

$$\frac{\mathrm{d}^2x}{\mathrm{d}t^2} + 0.1\frac{\mathrm{d}x}{\mathrm{d}t} + \sin x = 2\cos t.$$

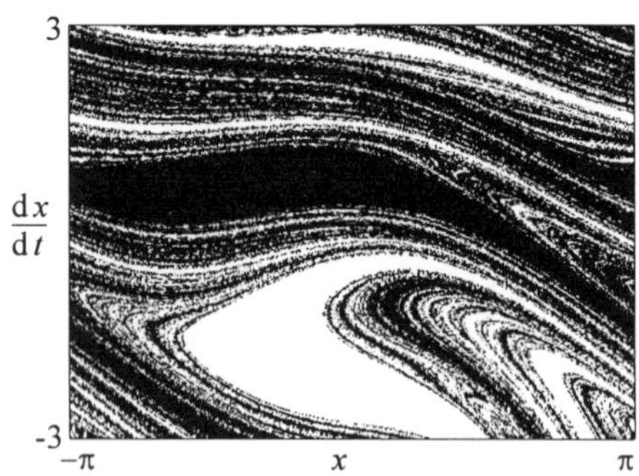


**Figure 11** Fractal basin boundary of a driven pendulum

In this case, there are two periodic attractors that have the same period as the forcing. The orbit for one of these attractors has an average clockwise motion (a negative average value of $\mathrm{d}x/\mathrm{d}t$), while the orbit for the other attractor has an average counter-clockwise motion.
In Figure 11, the black region represents initial values of $x$ and $\mathrm{d}x/\mathrm{d}t$ that tend to the attractor whose orbit has an average counter-clockwise motion, while the white region represents initial values that generate a clockwise motion. We can see that there is a small-scale structure on which the black and white regions appear to be finely interwoven, i.e. the basin boundary shows fractal nature.

*Reference*: McDonald *et al.* (1985).

**basin-boundary metamorphoses**
As control parameters of the dynamical system are varied, the character of a basin boundary can change. These changes are called basin-boundary metamorphoses.

*Reference*: Ott (1992).

**basin erosion**
After a homoclinic or heteroclinic tangency in driven oscillators, the basin of attraction for the locally stable solution becomes typically eroded as parameters vary, so that increasingly more and more conditions now lead to alternative co-existing solutions.

*Reference*: Thompson and Bishop (1994).

**The BDS test**
The Brock-Dechert-Scheinkman test is applied in economic modeling, which uses the correlation dimension to test the null hypothesis that a given time series is independently and identically distributed against a set of alternatives that includes determinism.

*Reference*: Grebogi and Yorke (1997).

**Belousov–Zhabotinskii reaction**
The Belousov–Zhabotinskii (BZ) reaction was originally proposed by Belousov when investigating how cells break down food. Belousov noticed that the chemical mixture alternated between a clear and coloured mixture but the discovery was not picked up until studied much later by Zhabotinskii. In addition to the periodic colour changes the reaction can display spiral

patterns. It is the best known example of a chemical system which exhibits both periodic and chaotic behaviour. The simplest mathematical model for the process can be represented as follows.

Consider the chemical reaction

$$A + B \overset{k_f}{\rightleftharpoons} C.$$

The reactants $A$ and $B$ are introduced into a closed container governed by a flow rate $r$, while an exit port allows excess material C to be removed. The rate equations

$$\frac{dA}{dt} = -k_f AB + k_d C - r(A - A_0)$$

$$\frac{dB}{dt} = k_f AB + k_r C - r(B - B_0)$$

and

$$\frac{dC}{dt} = k_f AB - k_r C - rC$$

where $k_d$ and $k_r$ are constants, exhibit nonlinear coupling between the chemical concentrations. $A_0$ and $B_0$ are the reactant concentrations at the input port ($C_0 = 0$). The experimental arrangement is shown in Figure 12.

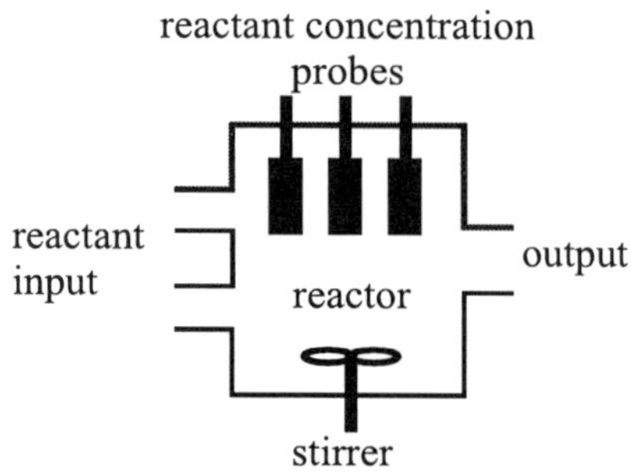


**Figure 12** Experimental arrangement of a chemical reaction with reactant flow; the probes monitor the reactant concentrations (adapted from Swinney (1983))

If $r$ is zero, then the reaction evolves to a state of equilibrium, while for large values of $r$ the materials are exhausted from the container before they have time to react. For intermediate values of $r$, the system has both periodic and chaotic states. The temperature-dependent rate constants and initial conditions also affect the dynamical state. Phase spaces can be constructed for the BZ reaction that allow the periodic and chaotic behaviour to be studied. Examples of these are shown in Figure 13.

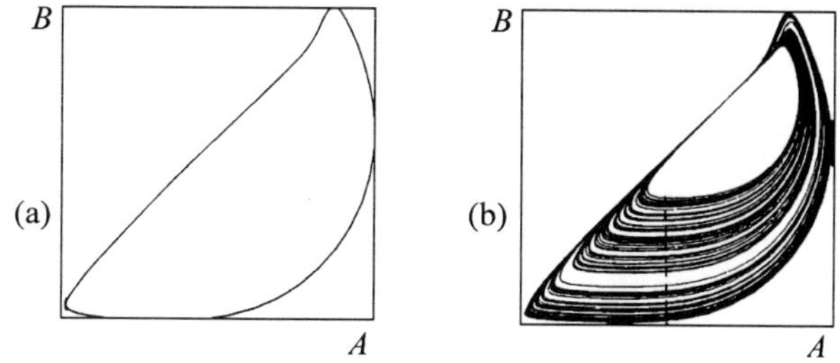


**Figure 13** Phase-space trajectories for the BZ reaction: (a) periodic behaviour; (b) chaotic behaviour

*Reference*: Swinney (1983).

### Bénard cells

Cells formed in a heated convecting fluid are called Bénard cells (*see* **Rayleigh–Bénard convection**).

### Bendixson's criterion

Consider a system of the coupled differential equations

$$\frac{\mathrm{d}x}{\mathrm{d}t} = f(x, y)$$

and

$$\frac{\mathrm{d}y}{\mathrm{d}t} = g(x, y)$$

where $(x, y) \in U \subseteq \mathcal{R}^2$, and in which $f$ and $g$ are sufficiently smooth. On a simply connected region $D \subseteq \mathcal{R}^2$, if the expression, $\partial f/\partial x + \partial g/\partial y$, is not identically zero and does not change sign, then the system has no closed orbits lying entirely in $D$.

*Reference*: Drazin (1992).

### Bernoulli shift map

The map $f : [0, 1) \to [0, 1)$ where $f(x) = 2x \bmod 1$ is called the Bernoulli shift map. More usually, we consider any full shift on $n$ symbols.

*Reference*: Jackson (1990).

### bifurcation

Consider a dynamical system

$$\frac{\mathrm{d}x}{\mathrm{d}t} = f(x, c)$$

where $x \in \mathcal{R}^n$, $c \in \mathcal{R}$ which has unique solutions $x(t, x_0, c)$. A bifurcation occurs when a change in the control (bifurcation) parameter, $c$, produces a change in the topology of the phase portrait. The partitioning of the control or parameter space, $c \in \mathcal{R}^n$, into regions where the phase portraits are topologically equivalent is called the control parameter space of the dynamical system. The boundary of these regions in the control parameter space is sometimes called the bifurcation set.

A plot of a variable representative of the dynamics (say $x$) versus the bifurcation parameter is referred to as a bifurcation diagram.

*Example 1*: Bifurcations of the system

$$\frac{\mathrm{d}x_1}{\mathrm{d}t} = -x_2$$

and

$$\frac{\mathrm{d}x_2}{\mathrm{d}t} = cx_1$$

subject to the condition $K(x, c') = x(t + c') - x(t) = 0$.

*Example 2*: Bifurcations of the fixed points of the one-dimensional map

$$x_{n+1} = f(x_n, c) \tag{1}$$

where $f$ is continuous and differentiable. Bifurcations of a period-$n$ orbit can be reduced to consideration of the map iterated $n$ times, map $f^n$, for which each point of the period $n$ orbit is a fixed point.
We shall consider only generic bifurcations, i.e. such bifurcations whose basic character cannot be altered by arbitrarily small continuous and differentiable perturbations. There are three generic types of bifurcations of continuous and differentiable one-dimensional maps:

(i) the supercritical (stable) period-doubling (flip) bifurcation;

(ii) the tangent bifurcation;

(iii) the subcritical period-doubling bifurcation.

These are illustrated in Figure 14 in the form of a bifurcation diagram. Dashed lines are used for solution paths corresponding to unstable orbits and solid lines for stable orbits. The parameter $c$ is assumed to increase to the right. Additionally in Figure 14, we define forward and backward bifurcations. In Figure 15, we show how the three forward bifurcations can occur as the shape of the map changes with an increase in the control parameter $c$.

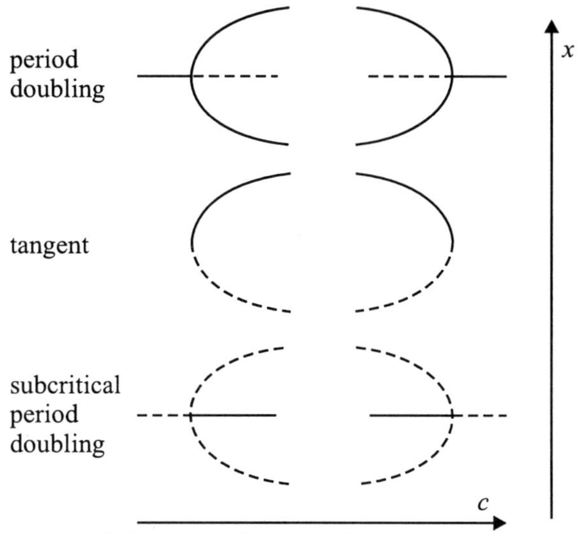


**Figure 14** Bifurcation diagrams for generic bifurcations of one-dimensional maps

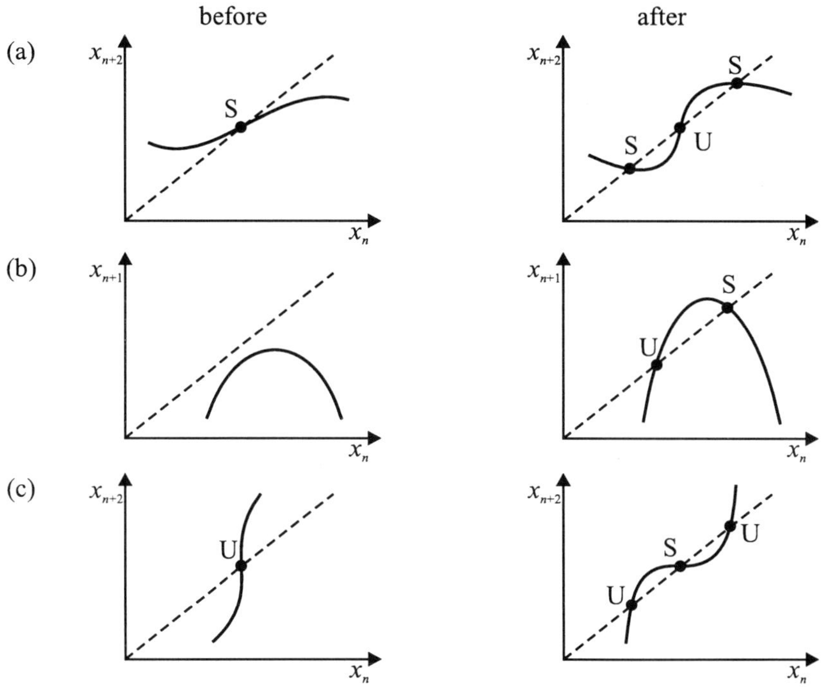


**Figure 15** Behaviour of a one-dimensional map before and after generic bifurcations: (a) supercritical period-doubling; (b) tangent; (c) subcritical period-doubling

*Example 3*: Local bifurcations of a fixed point $x = 0$ of a flow, together with examples of systems in which they occur, are shown in Figure 16.

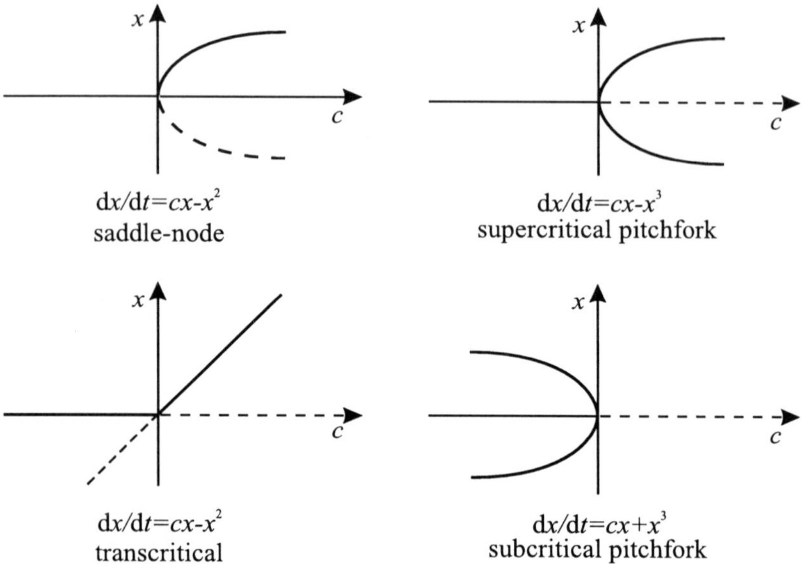


**Figure 16** Bifurcations of the fixed point

*Example 4*: Consider the dynamical system

$$\frac{\mathrm{d}x_1}{\mathrm{d}t} = c_1 + c_2 x_2 + x_2^2 + x_1 x_2$$

and

$$\frac{\mathrm{d}x_2}{\mathrm{d}t} = x_1.$$

The fixed points of this system are $x_1 = 0$, and $x_2 = -1/2c_2 \pm 1/2[c_2^2 - 4c_1]^{1/2}$, provided that $c_2^2 > 4c_1$. These fixed points are degenerate along the curve $c_2^2 = 4c_1$ in the parameter space.

If the system evolves so that $c_1$ or $c_2$ cross this curve, then there is a localised change in the phase space, namely two fixed points come together and vanish. From consideration of the Poincaré index (*see* **Poincaré index**) one finds that these simple fixed points must be a saddle and a node. The transition from a saddle point to a node is called a saddle–node bifurcation. In addition to the bifurcations determined by this local degeneracy of fixed points, it turns out that, in the region $c_2^2 > 4c_1$, there is both a curve associated with a Hopf bifurcation, and a curve associated with global bifurcation. These curves are shown in Figure 17. The bifurcation lines are labelled $(SN^-, SN^+, H, SC)$, which divide the control space into regions $(A, B, C, D)$. The local saddle–node and Hopf bifurcations occur across the lines $SN^\pm$ and $H$, respectively, whereas across the curve $SC$ a global bifurcation occurs which is called a saddle connection (homoclinic) bifurcation. The phase portraits corresponding to the above regions are shown in Figure 18 (*see also* **global bifurcation, Hopf bifurcation, Neimark bifurcation, period doubling bifurcation, symmetry–breaking bifurcation**).

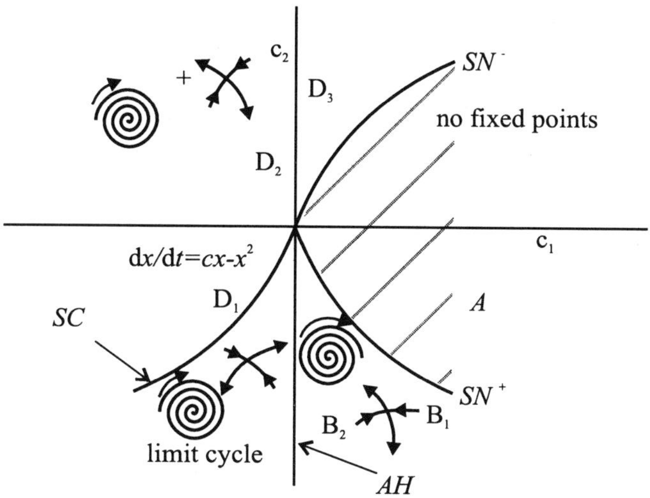


**Figure 17** Bifurcation diagram

*Remark*: The saddle–node bifurcation is also sometimes called a fold because of the nature of the response curve near the bifurcation.

*References*: Arnold (1984); Jackson (1990); Thom (1975); Thompson and Stewart (1986).

**bifurcation diagram** (*see* **bifurcation**)

**bifurcation parameter** (*see* **bifurcation**)

**bijection**
A one-to-one and onto map is called a bijection or a bijection mapping.

*Remark*: A bijection mapping can be discontinuous.

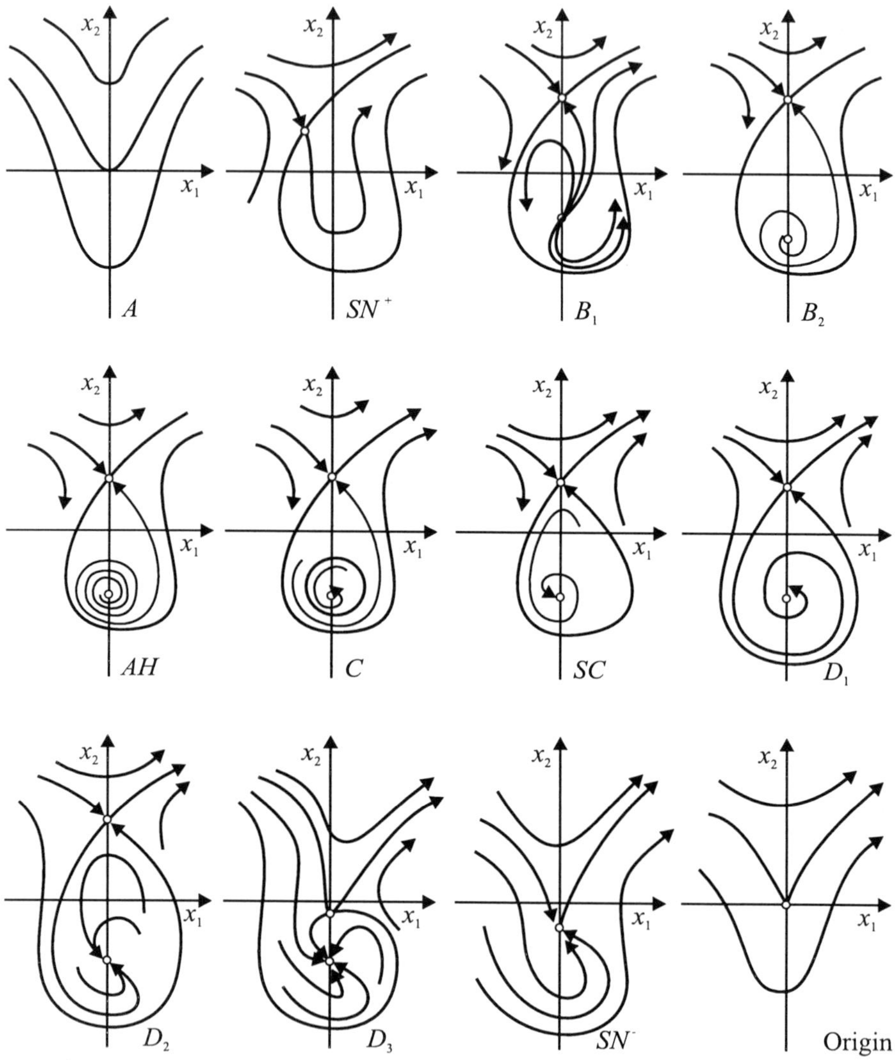


**Figure 18** Phase portraits

**billiard**
Let $Q$ be a connected domain in the plane $\mathcal{R}^2$ with a piecewise smooth boundary $\delta Q$. By a billiard in $Q$ we mean the dynamical system arising from the uniform motion of a point mass inside $Q$ with elastic reflections.

*Reference*: Berry (1989).

**Birkhoff ergodic theorem**
Let $(X, \mathcal{B}, m)$ be a probabilistic space. Suppose that the mapping

$$T : (X, \mathcal{B}, m) \to (X, \mathcal{B}, m)$$

is measure-preserving and $f \in L^1(m)$. Then

$$\frac{1}{n} \sum_{i=0}^{n-1} f(T^i(x))$$

converges almost everywhere to a function $f^* \in L^1(m)$. In addition,

$$f^* \circ T = f^*$$

almost everywhere, and if $m(X) < \infty$ then

$$\int f^* \mathrm{d}m = \int f \mathrm{d}m.$$

*References*: Arnold and Avez (1968); Steeb (1991).

**Birkhoff limit sets**
$\alpha$-limit sets and $\omega$-limit sets are called Birkhoff limit sets (*see* **alpha-limit set** and **omega-limit set**).

**Birkhoff–Shaw attractor**
The chaotic attractor of

$$\frac{\mathrm{d}x}{\mathrm{d}t} = 10x(0.1 - y^2)$$

and

$$\frac{\mathrm{d}y}{\mathrm{d}t} = 0.25 \sin 1.57t$$

which is equivalent to the velocity-forced van der Pol equation is called a Birkhoff–Shaw attractor.

*Reference*: Thompson and Stewart (1986).

**bistability**
If a dynamical system has two attractors then we say that the system is in a state of bistability. If there are more than two attractors we have multistability.

**Bloch theory** (*see* **Floquet theory**)

**blowout bifurcation**
Assume that the dynamical system

$$\frac{\mathrm{d}x}{\mathrm{d}t} = f(x, c)$$

where $x \in \mathcal{R}^n$ and $c \in \mathcal{R}$, has an attractor $A$ located on an $m$-dimensional ($m < n$) invariant subspace of the $n$-dimensional phase space for $c \leq c_0$. If for $c > c_0$, the attractor $A$ is replaced by the n-dimensional attractor $B$, then the bifurcation which occurs at $c_0$ is called a blowout bifurcation.

**blue-sky catastrophe** (*see* **crisis**)

**Bogdanov–Takens bifurcation** (*see* **Takens–Bogdanov bifurcation**)

**Borel sets**
Let $X$ be a topological space. All sets which belong to the $\sigma$-algebra (*see* **$\sigma$-algebra**) spanned by the all open subsets of the space $X$ are called Borel sets.

**bouncing ball**
An example of a mechanical system that leads to a nonlinear map is a ball bouncing on a vibrating table. The model, which includes two parameters, serves as a conceptually easy bridge between the discrete map and important three-and four-parameter experiments, such as the driven nonlinear oscillator and the dynamics of the forced pendulum.

Consider the model shown in Figure 19 in which an elastic ball is falling under gravitational force on to a sinusoidally vibrating table. The equations of motion of the table and of the ball are as follows:

$$\begin{aligned} A(t) &= A_0 \sin(\omega t), \\ Z(t) &= Z_0 + Vt - gt^2/2, \end{aligned} \tag{1}$$

where $Z_0$, $A_0$ and $\omega$ are the initial position of the ball, the amplitude and the frequency of the table vibration, respectively, and $g$ is the gravitational acceleration. The motion of the table $A(t)$ and the motion of the ball $Z(t)$ are constrained through the nonelastic impact (defined as $A(T) = Z(t)$) in which

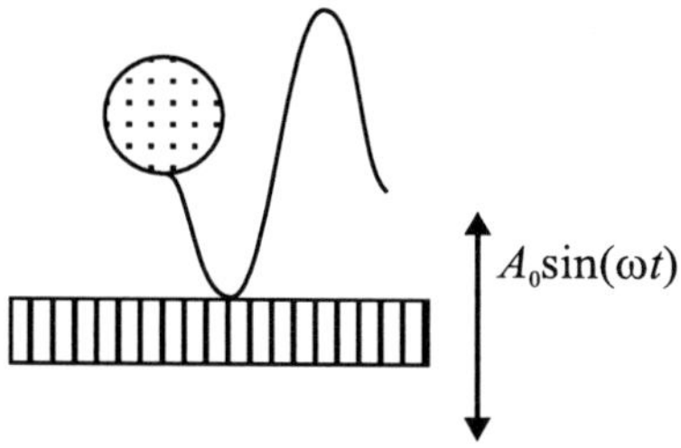


**Figure 19** Bouncing-ball model

$$K = -\frac{(V_n(t_n) - A_n(t_n))}{(U_n(t_n) - A_n(t_n))} \tag{2}$$

where $U_n$, $V_n$ and $A_n$ are, respectively, the absolute velocities of the approaching ball, the departing ball and the table, $K$ is the coefficient of restitution, and $t_n$ is the time of the $n$th impact. In some studies, it is assumed that the distance that the ball moves between impacts under the influence of gravity is large compared with the displacement of the table. In this case, the time interval between impacts can be approximated as

$$t_{n+1} - t_n = \frac{2V_n}{g} \tag{3}$$

and the velocity of approach at the $(n+1)$st impact is given by

$$U_n(t_{n+1}) = -V_n(t_n). \tag{4}$$

From equations (1)–(4), one obtains, after non-dimensionalising, the recurrence relationship between the state of the system at the $(n+1)$st and $n$th impacts in the form of a nonlinear map:

$$\begin{aligned} \phi_{n+1} &= \phi_n + v_n \\ v_{n+1} &= Kv_n - \delta\cos(\phi_n + v_n), \end{aligned} \tag{5}$$

where $\phi = \omega t$, $v = 2\omega V/g$ and $\delta = 2\omega^2(1+K)A_0/g$.

*Reference*: Holmes (1982).

**boundary**
The boundary $\partial A$ of a closed set $A$ is the set of points which belong to $A$, but which are not in the interior of $A$.

**boundary layer**
The boundary layer is a term used in fluid dynamics to represent the thin layer close to an object within a flow field at which a no-slip condition holds.

**bounded stability** (*see* **stability**)

**Bowen–Ruelle theorem (Sinai–Ruelle–Bowen theorem)**
For an Axiom-$A$ system, except for an initial set of Lebesgue measure zero, time averages exist for continuous phase functions.

More precisely, except for this initial set, a solution $x_t$ tends to some attractor $A$ as $t \to \infty$. The attractor $A$ has a canonical invariant ergodic measure called the SRB measure $\mu$, and if $\phi$ is a continuous function of the space, then

$$\lim_{T\to\infty} \frac{1}{T} \int_0^T \phi(x_t)\mathrm{d}t = \int_A \phi \mathrm{d}\mu.$$

*Remark*: This theorem applies to almost all solutions with respect to ordinary Lebesgue measure on the space, but the asymptotic behaviour is described by the measure $\mu$.

*References*: Eckmann and Ruelle (1985); Steeb (1991).

**box-counting dimension** (*see* **dimension of sets**)

**braid**
The projection of the periodic solutions of driven oscillators on to the $(x, t)$ plane in the interval of one period of the applied forcing produces a braid. Analysis of the crossings within a braid enables the solution to be classified and, moreover, the linking with other orbits and bifurcational precedences may be evaluated.

*References*: Ghrist and Holmes (1996); Thompson and Bishop (1994).

**branch point**
This is another term used for a bifurcation point, in particular for equilibria.

**Brouwer's fixed-point theorem**
For any continuous map $x_{n+1} = f(x_n)$ of the unit interval into itself, there must exist at least one fixed point.

*Reference*: Collet and Eckmann (1980).

**Brusselator equation**
The equations

$$\frac{\mathrm{d}x}{\mathrm{d}t} = a + x^2y - (1+b)x + A\cos\Omega t$$

and

$$\frac{\mathrm{d}y}{\mathrm{d}t} = bx - yx^2$$

where $a, b, A$ and $\Omega$ are constants, are called the Brusselator equations. These equations have their origin in the study of various chemical reactions carried out by a group at the Free University of Brussels.

*Reference*: Prigogine and Lefever (1968).

**Burger's equation**
The partial differential equation with nonlinearity and dissipation:

$$\frac{\partial \phi(x,t)}{\partial t} + (1 + \phi(x,t))\frac{\partial \phi(x,t)}{\partial x} - \frac{\partial^2 \phi(x,t)}{\partial x^2} = 0$$

is called the Burger's equation. This equation particularly arises in the theory of solitons.

**butterfly**
One of the elementary catastrophes (*see* **Thom's theorem**).

**butterfly effect**
It had been known for some time that certain dynamical systems were sensitive to initial conditions (e.g. consider the inverted position of a simple pendulum). H. Poincaré was certainly aware of this limitation for predictability and many researchers had noted this phenomenon in simulations (e.g. Y. Ueda has reported early chaotic attractors dating back to 1960). However, one of the earliest, and certainly most quoted computational demonstrations of sensitivity to initial conditions in a chaotic systems was given by Lorenz in his paper 'Does the flap of a butterfly's wings in Brazil set off a tornado in Texas'. Hence the phenomenon of sensitive dependence is often referred to as the *butterfly effect*. The fact that the chaotic attractor produced by Lorenz, when plotted in the phase space (*see* **Lorenz model**), has the shape of a butterfly, reinforced this as an appropriate name for this effect (*see* **sensitive dependence on initial conditions**).

*Reference*: Lorenz (1963).

# C

## Cantor set

A Cantor set $\Gamma$ is a closed set with the following properties:

(i) the largest connected subset of $\Gamma$ is a point;

(ii) every point of $\Gamma$ is a limit point of $\Gamma$.

*Example*: The triadic Cantor set is a subset of the line $\mathcal{R}$. A sequence of approximations is first defined as follows: start with the closed interval $C_0 = [0, 1]$, and the set $C_1$ is then obtained by removing the 'middle third' from $[0, 1]$, leaving $[0, 1/3] \cup [2/3, 1]$. The next set $C_2$ is defined by removing the middle third of each of the two intervals of $C_1$. This leaves

$$C_2 = [0, 1/9] \cup [2/9, 1/3] \cup [2/3, 7/9] \cup [8/9, 1]$$

and so on (*see* Figure 20). The triadic Cantor set is the 'limit' $C$ of the sequence $C_n$ of the sets. The sets decrease in the order: $C_0 \subseteq C_1 \subseteq C_2 \subseteq \cdots$, etc. and we define the 'limit' to be the intersection of the sets,

$$C = \bigcap_{k \in \mathcal{N}} C_k.$$

**Figure 20** The triadic Cantor set

**capacity dimension** (*see* **dimension of sets**)

**cardiac arrhythmias**
The normal heart rhythm is set by the sinoatrial node in the atria of the heart, but abnormal rhythms or arrythmias can occur, some of which can be life threatening. Over recent years there have been attempts to use a nonlinear dynamic approach to modelling heart rhythms as the behaviour is similar to that seen in coupled oscillators (*see* Figure 21).

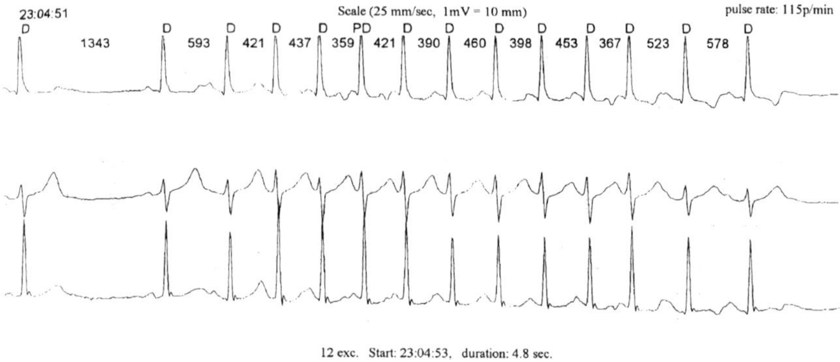


**Figure 21** Example of cardiac arrythmia — tachycardia

*Reference*: Grebogi and Yorke (1997).

**cat map**
Let $T^2 = \{(x, y) \bmod 1\}$ be the two-dimensional torus. The automorphism $f : T^2 \to T^2$ defined by

$$f(x, y) = (x + y, x + 2y) \bmod 1$$

is called the cat map. This map is also sometimes referred to as the Arnold cat map, after Arnold and Avez (1968) illustrated the dynamics of such a map by showing how a drawing of a cat's head was altered by $f^1$ and $f^2$, as seen in Figure 22.

*References*: Arnold and Avez(1968); Drazin (1992).

**catastrophe set** (*see* **bifurcation**)

**catastrophes**
Catastrophes are sudden changes representing discontinuous responses of

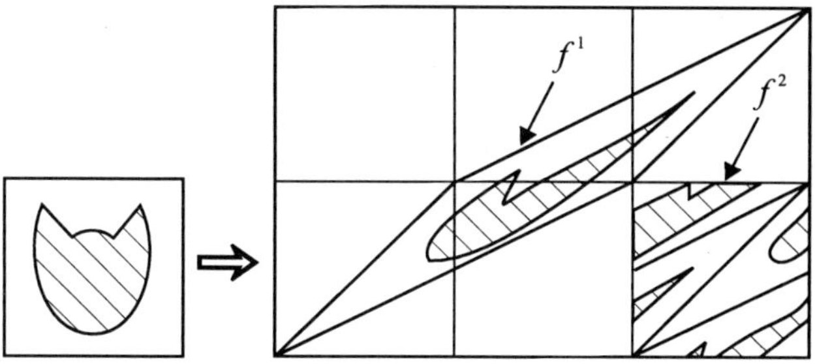


**Figure 22** Representation of a cat map

systems to smooth changes in the external conditions (*see* **bifurcation**). Catastrophe theory investigates and classifies these changes (*see also* **Thom's theorem**).

*Reference*: Arnold (1984).

**caustics**

Caustic is the term which arises in the use of a system of rays to describe the propagation of disturbances. The propagation of a disturbance inside an ellipse can be described by using the family of internal normals to the ellipse, as shown in Figure 23.

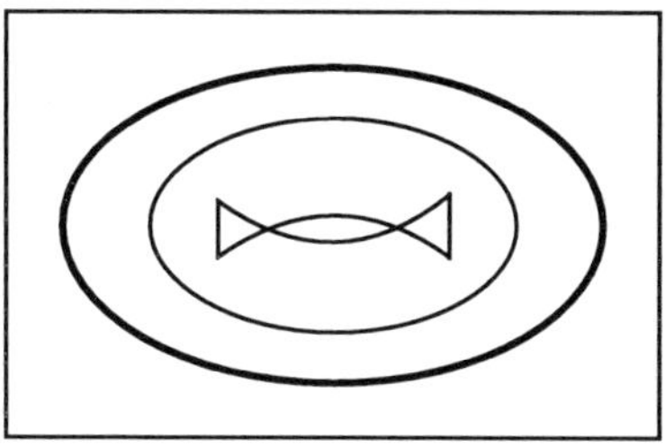

**Figure 23** Propagation of disturbances inside an ellipse

The family of normals has an envelope which is called a caustic.

*Example*: A rainbow in the sky is due to the caustic of the system of rays that have been completely reflected by water droplets.

*Reference*: Arnold (1984).

**celestial mechanics**

The investigation of the motions of stars and planets is referred to as celestial mechanics. In recent years, there has been considerable renewed interest in celestial mechanics since the use of high-speed computers allows the long-term integration of equations which govern planetary motion. Of particular interest for nonlinear dynamics is the chaotic behaviour of hyperion, and questions relating to whether the solar system is stable or not; notable work concerning the latter has been carried out by Lasker.

*References*: Dankowicz (1997); Laskar (1989), Moser (1973), Sussman and Wisdom (1992).

**cell map**

A cell map is a plot in initial-condition space of the basins of attraction for the various co-existing solutions at fixed parameter values. A crude way of determining the basins is to consider a portion of the phase space and then divide it into a small grid so that each grid point defines an initial condition. Each grid point can then be integrated forward to determine the subsequent long-term behaviour. While simple and effective, this ('carpet-bombing') approach is numerically inefficient. The cell mapping method, or cell-to-cell mapping method, was devised by Hsu to minimise the necessary computations. By considering cells (portions of the phase space), rather than grid points, a cell is characterised by its central point which may be integrated forward in time as before, the subsequent attracting solution noted, and all cells visited during the evolution (rather than just the one point) contoured since they represent initial conditions which would lead to the steady-state solution. If the evolution of a further cell at some time coincides with a cell already labelled as leading to an attractor, then the evolution routine can be terminated and all cells labelled accordingly as before. An example of a cell map for the escape equation (*see* **escape equation**) is shown in Figure 24. Different shading corresponds to initial conditions which converge to different attracting solutions, i.e. pale grey represents an escape to infinity, while dark grey and black represent solutions within the well.

*Reference*: Hsu (1987).

**cellular automata**

A cellular automata (CA) is a spatially extended dynamical system with discrete space, discrete time and discrete state space. If the state space is continuous, this leads to a coupled map lattice (*see* **coupled map lattice**). Consider a discrete lattice $\mathcal{L}$ and let $\mathcal{S}$ be a discrete, finite set of possible states. Let $x_i(t) \in \mathcal{S}$ be a dynamical variable at site $i \in \mathcal{L}$ and consider $N(i)$ to be a finite neighbourhood of $i$ in $\mathcal{L}$. A cellular automata maps $\{x_j(t)\}_{j\in N(i)}$ into $x_i(t+1)$.

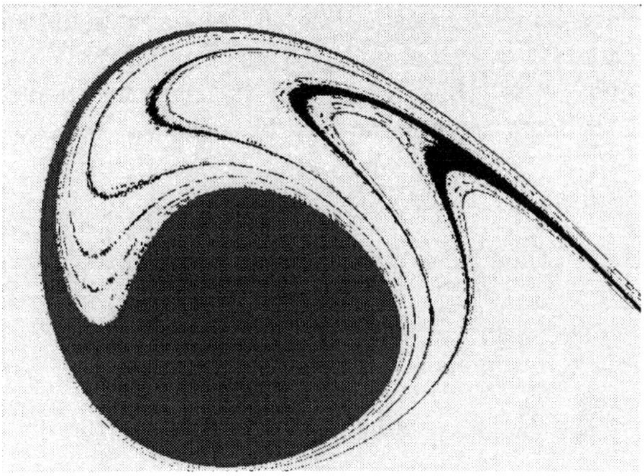

**Figure 24** A cell map for the escape equation

Cellular automata are usually classified by their asymptotic behaviour as follows:

Class I, fixed homogeneous states;

Class II, periodic patterns;

Class III, chaotic aperiodic patterns;

Class IV, complex spatio-temporal patterns.

*Reference*: Wolfram (1986).

**centre (center)** (*see* **fixed points**)

**centre manifold theorem**
The centre manifold theorem provides a means for systematically reducing the dimension of the phase space which needs to be considered when analysing bifurcation problems.

Let $f$ be a $C^r$ vector field on $\mathcal{R}^n$, vanishing at the origin, i.e. $f(0) = 0$, and let $A$ be the Jacobian matrix of $f(0)$. Divide the eigenvalues of $A$ into three parts $\sigma_s, \sigma_c, \sigma_u$ with

$$\text{Re } \lambda : \begin{cases} < 0 & \text{if} \quad \lambda \in \sigma_s \\ = 0 & \text{if} \quad \lambda \in \sigma_c \\ > 0 & \text{if} \quad \lambda \in \sigma_u \end{cases}$$

Let the generalised eigenspaces of $\sigma_s, \sigma_c$, and $\sigma_u$ be $E^s, E^c$, and $E^u$, respectively. Then there exist $C^r$ stable and unstable manifolds, $W^s$ and $W^u$, tangent to $E^s$ and $E^u$ at 0, and a $C^{r-1}$ centre manifold $W^c$, tangent to $E^c$ at 0. The manifolds $W^s, W^u$ and $W^c$ are all invariant for the flow $f$. The stable and unstable manifolds are unique, but $W^c$ need not necessarily be unique.

*Remarks*:

1. If $f$ is $C^\infty$, then we can find a $C^r$ centre manifold for any $r < \infty$.

2. In the control-phase space the surface which contains only periodic solutions is a centre manifold (*see* **Hopf bifurcation**).

*References*: Carr (1981); Guckenheimer and Holmes (1983).

**chaos**

Chaos is a technical expression for a specific type of irregular motion produced by a deterministic system. Chaos is a long term phenomenon and therefore when energy dissipation occurs, for example in mechanical systems, then a continuous input of energy is required to maintain the chaotic response otherwise any viewed irregularity will only be transient. Chaos has now been viewed in a wide variety of physical systems including mechanical, fluid, electronic, chemical and even biological experiments. Chaos lies within a well ordered structure and as such is not chaotic in the everyday sense of the word. To describe this type of motion, Ueda used the expression "randomly transitional phenomena" but following its usage by Li and Yorke the word chaos is now firmly fixed as the term used.

More mathematically, consider a typical orbit $\gamma$ evolving on the attractor of the dynamical system

$$\frac{\mathrm{d}x}{\mathrm{d}t} = f(x)$$

where $x \in \mathcal{R}^n$. The orbit $\gamma$ is chaotic when $\gamma$ is characterised by at least one positive Lyapunov exponent, while the sum of all Lyapunov exponents is less or equal to zero (less refers to the gradient, and equal to conservative systems).

A map $f : S \to S$, where $S \subset \mathcal{R}^n$, is chaotic if at least one Lyapunov exponent if positive.

Let $X$ be a set. A map $f : X \to X$ is said to be chaotic on $X$ if:

(i) $f$ has sensitive dependence on initial conditions, i.e. there exists $\delta > 0$ such that, for any $x \in X$ and any neighbourhood $U$ of $x$, there exists $y \in X$ and $n \geq 0$ such that $|f^n(x) - f^n(y)| > \delta$;

(ii) $f$ is topologically transitive, i.e. for any pair of open sets $V, W \subset X$ there exists $k > 0$ such that $f^k(V) \cap W \neq \emptyset$;

(iii) periodic points are dense in $X$.

*References*: Kapitaniak (1991); Li and Yorke (1975); Ueda (1979); Steeb (1991).

**chaos–chaos intermittency** (*see* **crisis, intermittency**)

**chaotic saddle**
A chaotic invariant set $A$ is a chaotic saddle if there is a neighbourhood $U$ of $A$ such that $b(A) \cap U$, where $b(A)$ is the basin of attraction of $A$, is greater than $A$ but has zero Lebesque measure.

*Remark*: A saddle fixed point also satisfies this definition.

$A$ is a normally repelling chaotic saddle if it is an attractor in the invariant subspace, but all points not lying on this subspace eventually leave a neighbourhood of $A$.

*Reference*: Ashwin *et al.* (1994).

**chaotic scattering**
Consider a classical scattering problem for a conservative dynamical system, i.e. the motion without friction of a point particle in a potential $V(x)$ for which $V(x)$ is zero outside some finite region of the space which is called the scattering region. Thus the particle moves along straight lines sufficiently far outside the scattering region. We envisage that a particle moves towards the scattering region from outside, interacts with the scatterer, and then leaves the scattering region. The main problem is the dependence of the motion far from the scatterer on the motion close to the scatterer, before scattering takes place. The transition from regular to chaotic scattering occurs via an abrupt or massive bifurcation.

*Example*: Consider the example given in Figure 25, which shows a scattering problem in two dimensions. The incident particle has a velocity parallel to the $x$-axis at a vertical displacement $y = b$. After interaction with the scatterer, the particle moves off to infinity, with its velocity vector making an angle $\phi$ to the $x$-axis. The quantities $b$ and $\phi$ are termed the impact parameter and scattering angle, respectively.

Now consider the functional dependence of $\phi$ on $b$. If the function $\phi(b)$ is

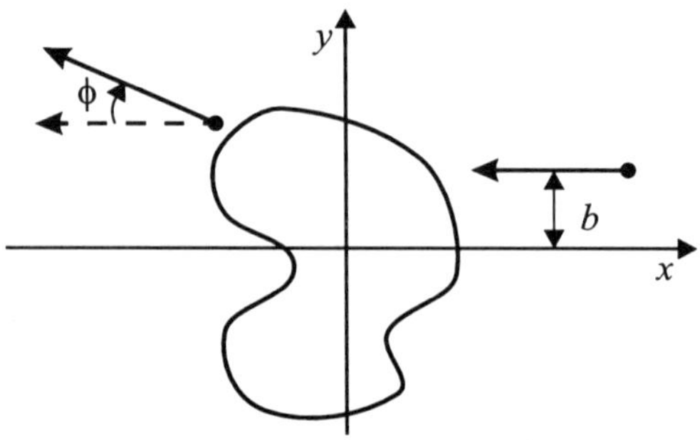


**Figure 25** Scattering in two dimensions

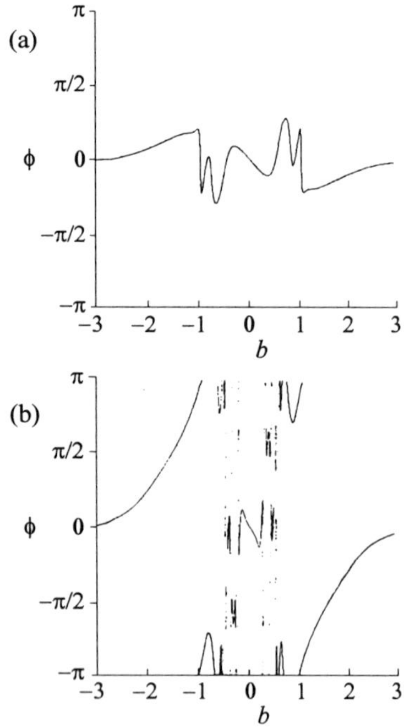


**Figure 26** Illustrations of (a) regular and (b) chaotic scattering

continuous, e.g. as shown in Figure 26(a), then we have regular scattering. However, if this function is of the type shown in Figure 26(b), we refer to the behaviour as chaotic scattering.

*Reference*: Ott (1992).

**characteristic function**
Let $A \subset \mathcal{R}^n$; then a function

$$f_A = \begin{cases} 1 & \text{for} \quad x \in A \\ 0 & \text{for} \quad x \notin A \end{cases}$$

is called the characteristic function of the set $A$.

**characteristic multipliers (number)** (*see* **Floquet theory**)

**Chua's circuit**
The electronic circuit which is shown in Figure 27 is known as a Double-Scroll or Chua's circuit. This is a third-order circuit which has only one nonlinear element, namely a piecewise linear resistor (see Figure 28).

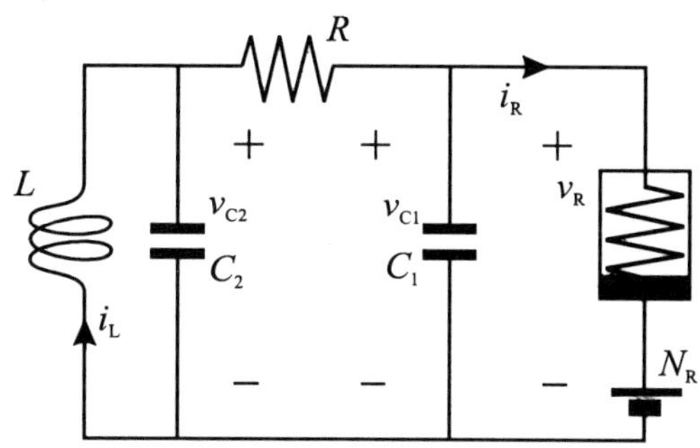


**Figure 27** Chua's (double-scroll) circuit

The state equations of Chua's circuit are as follows:

$$C_1 \frac{\mathrm{d}v_{C_1}}{\mathrm{d}t} = G(v_{C_2} - v_{C_1}) - f(v_{C_1}) \tag{1a}$$

$$C_2 \frac{\mathrm{d}v_{C_2}}{\mathrm{d}t} = G(v_{C_1} - v_{C_2}) + i_L \tag{1b}$$

and

$$L \frac{\mathrm{d}i_L}{\mathrm{d}t} = v_{C_2} \tag{1c}$$

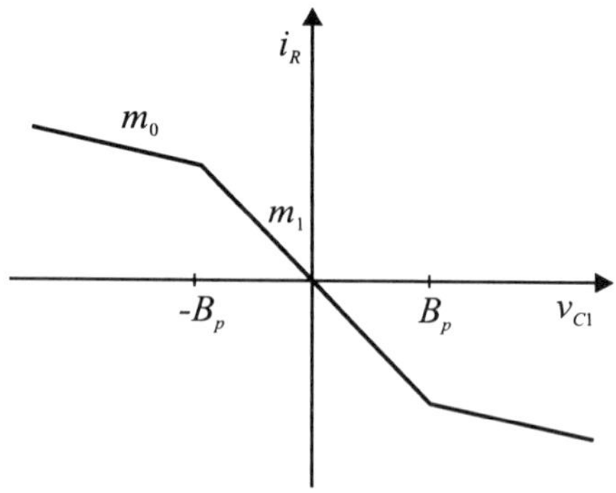


**Figure 28** Characteristic form of the nonlinear element

where $G = 1/R$, and a three-segment piecewise linear $v_{C_1} - i$, characteristic of the nonlinear element, is defined by

$$f(v_{C_1}) = m_0 v_{C_1} + \frac{1}{2}(m_1 - m_0)[|v_{C_1} + B_p| - |v_{C_1} - B_p|].$$

This relationship is shown graphically in Figure 28; where the slopes in the inner and outer regions are $m_1$ and $m_0$, respectively, and $\pm B_p$ denotes the break points.

Equations (1a–c) can be rewritten in the dimensionless forms

$$\dot{x} = \alpha(y - x - f(x)) \tag{2a}$$

$$\dot{y} = x - y + z \tag{2b}$$

and

$$\dot{z} = -\beta y \tag{2c}$$

where $x = v_{C_1}/B_p$, $y = v_{C_2}/B_p$, $z = i/B_p G$, $\alpha = C_2/C_1$, $\beta = C_2/G^2 L$, $f(x) = bx + 1/2(a - b)[|x + 1| - |x - 1|]$, $a = m_1/G$, and $b = m_0/G$.

Chua's circuit is one of the simplest electronic circuits which can display chaos and possibly the most studied in this sense. Chaos cannot occur in an autonomous circuit (modelled by nonlinear-state equations) with fewer than three energy storage elements (capacitors and inductors), and at least one nonlinear active element is needed even for oscillation to be possible.

*Example*: For the conditions $\alpha = 10$, $\beta = 14.87$, $a = -1.27$ and $b = -0.68$, Chua's circuit operates on the chaotic double-scroll attractor shown in Figure 29.

*References*: Chua (1992); Matsumoto (1984).

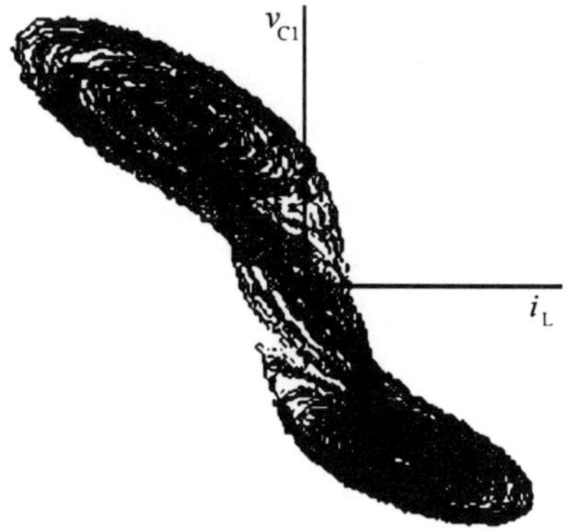


**Figure 29** Double scroll attractor

**circle map**
A map defined as

$$\phi_{n+1} = \phi_n + 2\pi K + V \sin \phi_n \bmod 2\pi,$$

where $K$ and $V$ are real parameters, is called a circle map.

*Reference*: Schuster (1984).

**$C^k$ function**
A function is $C^k$ ($k = 0, 1, 2, ..., \infty$) if it is $k$-times differentiable and its $k$th derivative is continuous.

**closed orbits**
Closed orbits in phase space correspond to periodic solutions.

**closed set**
The set $A$ is closed if it contains all of its limit points, i.e. if $x_n$ is a sequence in $A$ converging to a point $x$, then $x$ is in $A$.

**closure**
The closure $\bar{A}$ of a set $A$ is the union of $A$ and its set of limit points.

**codimension**
The codimension of a $k$-dimensional submanifold of an $n$-dimensional manifold is $n - k$.

The codimension of a bifurcation is the number of parameters that have to be specified in order to define the bifurcation.

*Remark*: The knowledge of the codimension of the bifurcation is very useful in the further analysis. Usually the considered system is embedded in a parameterized family of systems transverse to the bifurcation surface with the number of parameters equal to codimension of the bifurcation. These parameterized systems are called unfoldings and, if they contain all possible qualitative dynamics that can occur near the bifurcation, they are called universal unfoldings.

*Reference*: Wiggins (1991).

**co-existing**

Dynamical systems in general and nonlinear systems in particular may typically have co-existing solutions at certain fixed parameter values.

*Example*: The simplest case is possibly the resonant and non-resonant motions of a driven oscillator due to a nonlinear restoring force. For example, consider a driven oscillator with a softening spring force $f(x)$ given by

$$\frac{\mathrm{d}^2x}{\mathrm{d}t^2} + b\frac{\mathrm{d}x}{\mathrm{d}t} + f(x) = \cos \omega t.$$

The amplitude response curve of period-one motions bends to the left as shown in Figure 30.

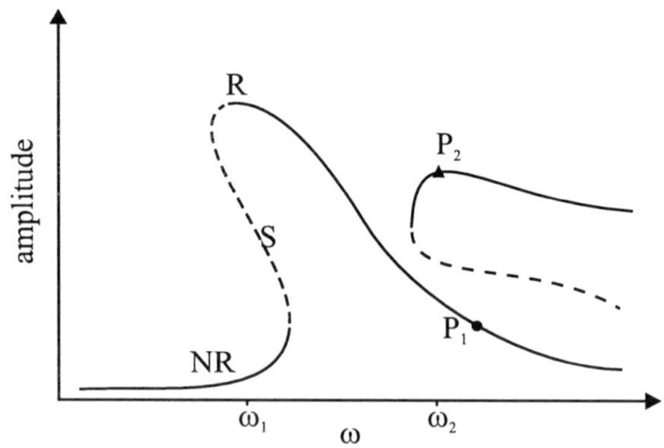


**Figure 30** Nonlinear softening amplitude response curve

During the hysteresis, say for example near $\omega = \omega_1$, there exist two stable solutions, namely a large-amplitude resonant motion (R), and a small-amplitude nonresonant motion (NR), together with an unstable saddle (S) solution with the same frequency.

Alternatively, at some parameter value, say $\omega = \omega_2$, a period-two solution ($P_2$) may co-exist with the harmonic period-one solution ($P_1$).

**collision of attractors or basins** (*see* **crisis**)

**compact set**
The set $A$ is compact if and only every open cover has a finite subcover, and vice versa.

*Remark*: For subsets of Euclidean space, any compact set is closed and bounded.

**compact support**
A function $f : \mathcal{R}^n \to \mathcal{R}$ has compact support if the set of points $x$, such that $f(x) \neq 0$, is bounded.

**complexity**
Complexity is the term which refers to all of the science of complex adaptive systems.

**conjugate** (*see* **topological conjugate**)

**Conley–Moser conditions**
The necessary conditions to prove that a two-dimensional invertible map $f : \mathcal{R}^n \to \mathcal{R}^n$ has an invariant Cantor set on which the dynamic is topologically conjugate to a full shift on $N$ symbols, where $N \geq 2$, are called Conley–Moser conditions.

*Reference*: Wiggins (1990).

**connected**
A set $S$ is not connected if there exist two open sets $U_1$ and $U_2$ such that

(i) $U_1 \cap \bar{U}_j = \emptyset$ for $i \neq j$;

(ii) $U_i \cap S \neq \emptyset$ for $i = 1, 2$;

(iii) $S \subset U_1 \cup U_2$.

**conservative system**
If we choose, for $t = 0$, an initial closed $(n-1)$-dimensional surface $S_0$ enclosing a volume $V(0)$ in $n$-dimensional phase space, and then evolve each point on

the surface $S_0$ forward in time by using these as initial conditions of the dynamical system

$$\frac{\mathrm{d}x}{\mathrm{d}t} = f(x)$$

where $x \in \mathcal{R}^n$, then the closed surface $S_0$ evolves to a closed surface $S_t$, enclosing a volume $V(t)$ at some later time $t$. If for all $t$, $V(t) = V(0)$ then the dynamical system is said to be conservative (*see also* **Hamiltonian system**).

Conservative systems are measure preserving (*see* **measure–preserving**).

*Reference*: Ott (1992).

**constants of motion**
Consider an autonomous system

$$\frac{\mathrm{d}x}{\mathrm{d}t} = f(x, c) \tag{1}$$

where $x, c \in \mathcal{R}^n$. The general solution of this equation is of the form

$$x(t) = X(t, x_0) \tag{2}$$

where $x(0) = X(0, x_0) = x_0$ is an initial condition. Equation (2) can be viewed as a mapping in the phase space $X^t : x_0 \to x$, which carries the initial point $x_0$ to the point $x(t)$. If the dynamical solution exists and is unique, then for any given $x$ and $t$ there is a unique initial point $x_0$, i.e. one can write

$$x_0 = K(x, t). \tag{3}$$

The functions $K(x, t)$ in (3) are the $n$ constants of the motion, since the initial conditions $x_0$ are obviously constant, so for all $t' > t$

$$K(x(t), t) = K(x(t'), t'). \tag{4}$$

Any function $K(x(t), t)$ which satisfies (4), when $x(t)$ satisfies (1), is called a constant or integral of motion.

*References*: Jackson (1990), Steeb (1991).

**control parameter** (*see* **bifurcation**)

**controlling chaos**
This is a procedure whereby a systematic control algorithm is applied to replace the chaotic evolution of a dynamical system by a more desirable periodic or steady-state response.

We can divide strategies for the chaos control into two broad categories: first those in which the actual trajectory in the phase space of the system is monitored, and some feedback process is employed to stabilise the trajectory on to a desired mode, and secondly, non-feedback methods in which some other property or knowledge of the system is used to modify or exploit the chaotic behaviour.

Ott, Grebogi and Yorke have, in an important series of papers, proposed and developed the method by which chaos can always be suppressed by shadowing one of the infinitely many unstable periodic orbits (or perhaps steady states) embedded in the chaotic attractor. The basic assumptions of this method are as follows:

(i) The dynamics of the system can be described by an $n$-dimensional map of the form

$$\zeta_{n+1} = f(\zeta_n, p). \tag{1}$$

This map in the case of continuous-time systems can be constructed, e.g. by introducing a Poincaré map.

(ii) The variable $p$ is some accessible system parameter which can be changed in some small neighbourhood of its nominal value $p^*$ at which the system behaves chaotically.

(iii) For the value of $p^*$, there is a periodic orbit within the attractor around which we would like to stabilise the system.

(iv) The position of this orbit varies smoothly with small changes in the parameter $p$.

Let $x_F$ be a chosen fixed point of the map $f$ of the system existing for the parameter value $p^*$. In the close neighbourhood of this fixed point we can assume with good accuracy that the dynamic is linear and can be expressed approximately by

$$\zeta_{n+1} - \zeta_F = M(\zeta_n - \zeta_F). \tag{2}$$

The elements of the matrix $M$ can be calculated by using the measured chaotic time series and analysing its behaviour in the neighbourhood of the fixed point. Furthermore, the respective stable and unstable eigenvalues, $e_s$ and $e_u$, and the eigenvectors, $v_s$ and $v_u$, of this matrix can be found; the latter determine the stable and unstable manifolds in the neighbourhood of the fixed point.

Denoting by $f_s$ and $f_u$, the respective contravariant eigenvectors

$$f_s v_s = f_u v_u = 1$$

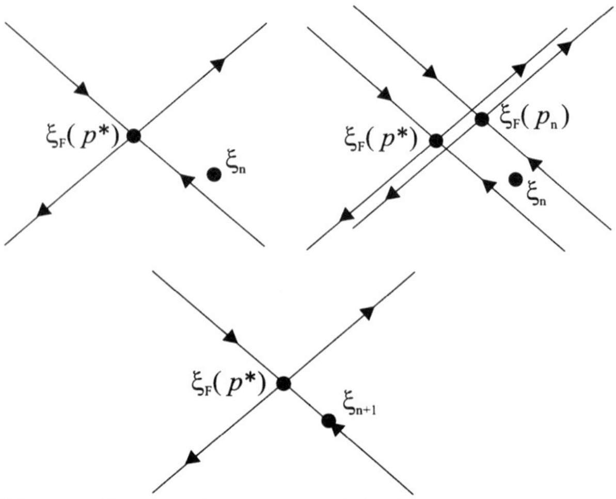


**Figure 31** An illustration of the basic concept of the Ott–Grebogi–Yorke method

and

$$f_s v_u = f_u v_s = 0,$$

we can find the linear approximation which is valid for small $|p_n - p^*|$; namely

$$\zeta_{n+1} = p_n g + [e_u v_u f_u + e_s v_s f_s][\zeta_n - p_n g|$$

where

$$g = \frac{\delta \zeta_f(p)}{\delta p}|_{p+p^*}.$$

The main idea of this method is to choose $p_n$ so that $\zeta_{n+1}$ should fall on the stable manifold of $\zeta_F$ such that $f_u \zeta_{n+1} = 0$:

$$p_n = \frac{e_u \zeta_n f_u}{(e_u - 1) g f_u}. \tag{3}$$

The concept of the Ott–Grebogi–Yorke (OGY) algorithm is schematically represented in Figure .

The main properties of the method are as follows:

(i) it is a feedback method;

(ii) any accessible system parameter can be used as a control parameter;

(iii) noise can destabilise the controlled orbit, thus resulting in occasional chaotic bursts;

(iv) Before approaching the desired periodic orbit, the trajectory exhibits a long chaotic transient.

*Example*: As an example of the application of this method, consider the control of chaos in Chua's circuit (*see* **Chua's circuit**) operating on the single scroll chaotic attractor. The block diagram of the implemented system is shown in Figure 32.

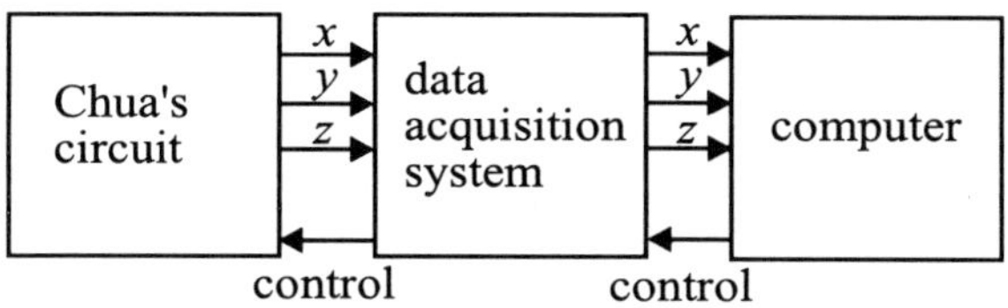


**Figure 32** A practical implementation of the Ott–Grebogi–Yorke method

Figure 33 shows the stabilisation of period-one and period-two unstable periodic orbits (denoted by black curves). Before the control is achieved, the trajectories exhibit chaotic transients (shown in grey).

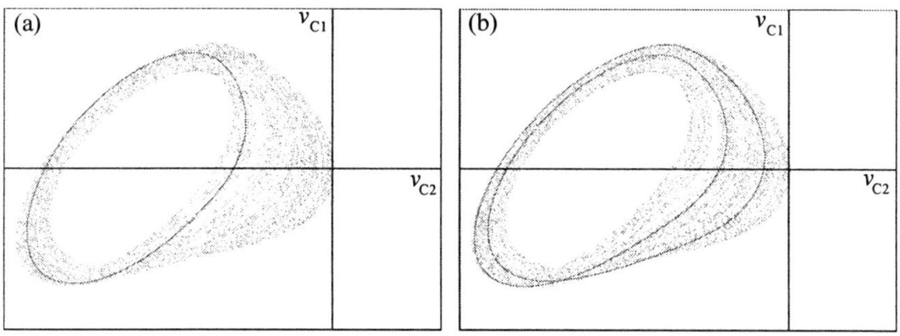


**Figure 33** Control of (a) period-one and (b) period-two orbits

The OGY approach has stimulated a great deal of research activity, including both theoretical and experimental studies (Kapitaniak 1996). The efficiency of the technique was demonstrated by Ditto *et al.* (1991) in a periodically forced system, converting its chaotic behaviour into period-one and period-two orbits, and the application of the method to stabilise higher-periodic orbits in a chaotic diode resonator was demonstrated by Hunt (1991). Another interesting application of the method is the generation of a desired aperiodic orbit – Mehta and Henderson (1991), while Tel (1991) was able to demonstrate controlled transient chaos. Related work by Dressler and Nitsche (1992) used time-delay techniques to control chaos.

*References*: Kapitaniak (1996); Ott *et al.* (1990).

**correlation dimension** (*see* **dimension of sets**)

**correlation integral** (*see* **correlation dimension**)

**cosine map**
The one-dimensional map

$$x_{n+1} = \cos x_n$$

where $x \in \mathcal{R}$, is called the cosine map.

**Couette–Taylor flow** (*see* **Taylor–Couette flow**)

**Coulomb friction** (*see* **damping**)

**coupled map lattice**
A coupled map lattice (CML) is a spatially extended dynamical system with discrete space, discrete time and continuous state space. If the state space is discrete it yields to a cellular automata (*see* **cellular automata**). Consider a discrete lattice $\mathcal{L}$, let $x_i(t) \in \mathcal{R}^n$ be the dynamical variable at site $i \in \mathcal{L}$ and consider $N(i)$ to be a finite neighbourhood of $i$. A CML maps $\{x_j(t)\}_{j \in N(i)}$ into $x_i(t+1)$.

The most widespread one-dimensional model of a CML is the so-called diffusive CML:

$$x_i(t+1) = (1-\epsilon)f(x_i(t)) + \frac{\epsilon}{2}(f(x_{i-1}(t)) + f(x_{i+1}(t)))$$

where $\epsilon \in [0,1]$ is the coupling parameter.

*Reference*: Kaneko (1993).

**coupled systems**
Two dynamical systems (A) and (B) described by the following equation

$$\frac{dX}{dt} = F(X,Y) \qquad \text{(A)}$$

$$\frac{dY}{dt} = G(X,Y) \qquad \text{(B)}$$

where $X, Y \in \mathcal{R}^n$, are coupled if functions $F$ and $G$ depend on both $X$ and $Y$.

If $F = F(X)$ and $G = G(Y)$, then the systems (A) and (B) are uncoupled.

If $G = G(Y)$, but $F = F(X, Y)$, we say that the systems (A) and (B) are unidirectionally coupled. There is no feedback and such systems are also referred to as drive-response systems.

**crisis**
Sudden qualitative changes of an attractor, which occur at bifurcation parameter values where the attractor collides with an unstable orbit, are called crises.

If the sudden destruction of a chaotic attractor occurs when the attractor collides with a periodic orbit on its basin boundary then the crisis is called a boundary crisis or blue-sky catastrophe.

If the sudden increase in the size of a chaotic attractor occurs when the periodic orbit with which the chaotic attractor collides is in the interior of its basin, this it is called an interior crisis.

In an attractor-merging crisis two or more chaotic attractors simultaneously collide with a periodic orbit or orbits on the basin boundary which separates them.

*Remark*: After a boundary crisis, we observe transient chaos.

Following an interior crisis or an attractor-merging crisis, we observe a crisis induced intermittency. This type of intermittency is characterised by permanent jumps between two chaotic attractors. It is also called chaos–chaos intermittency.

*References*: Grebogi *et al.* (1982), Thompson and Stewart (1986).

**critical damping** (*see* **damping**)

**critical points** (*see* **fixed points**)

**cubic map**
A map $[0,1] \to [0,1]$, given by

$$x_{n+1} = rx_n^3 + (1-r)x_n$$

where $r \in \mathcal{R}$, is called a cubic map.

**cusp**
One of the elementary catastrophes (*see* **Thom's theorem**).

**cusped-diamond**
One of the elementary catastrophes (*see* **Thom's theorem**).

**cycle**
This is a term which is sometimes used to represent a periodic orbit or limit cycle, typically in autonomous systems.

# D

**damping**

When considering structural systems, positive damping is the mechanism which models the dissipation of energy which results in a reduction in the amplitude of motion. In some cases, for instance in the van der Pol equation, the damping is negative, thus leading to an increase in the amplitude of motion. Damping typically occurs due to interactions between the system and its surroundings (e.g. via the damped base in a vibrating beam or interaction with the surrounding fluid — in this case, air) but may also be due to internal interactions (*see also* **van der Pol equation**).

Damping also occurs due to the interaction between a moving body and the surface on which it moves. When the contact between the two bodies is dry the force $F$ which opposes the motion is caused by microscopic irregularities of the sliding surfaces. Often in this situation the friction force is called Coulomb damping. When an external force $P$ is applied to the mass (as in Figure 34) the friction force increases until it reaches a critical value $F_0(=\mu N)$, at which stage the mass begins to move. The term $\mu$ is called the static coefficient of friction and $N$ is the normal force acting between the mass and the surface (*see also* **stick-slip systems**).

If the surfaces are separated by a thin film of lubricant then the friction force becomes fluid in nature and the term viscous damping is used. One

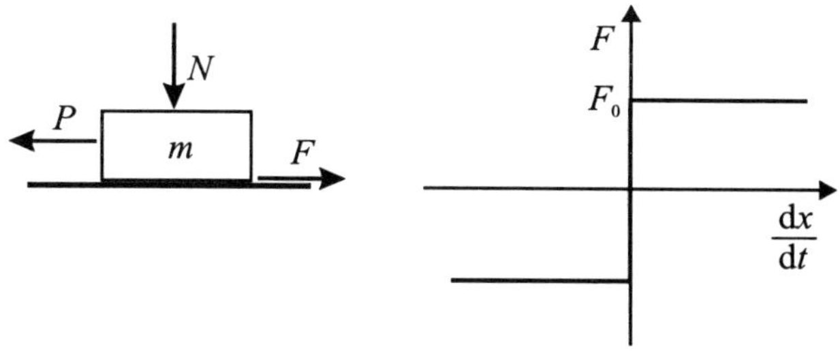


**Figure 34** Dry friction force (Coulomb damping)

approximation is to consider the force to be proportional to the velocity, i.e. $F = b\mathrm{d}x/\mathrm{d}t$, representing a so-called linear viscous damping force.

If an immersed body moves through a fluid at high Reynolds numbers then the flow can separate and the drag force is approximately governed by the velocity squared

$$F = b\frac{\mathrm{d}^2x}{\mathrm{d}t^2}.$$

*Critical damping*:

Consider the oscillator

$$\frac{\mathrm{d}^2x}{\mathrm{d}t^2} + 2\zeta\frac{\mathrm{d}x}{\mathrm{d}t} + x = 0.$$

For $\zeta < 1$ in the phase plane, we have a spiral evolution of transients towards the stable fixed point $\mathrm{d}x/\mathrm{d}t = x = 0$ in the phase plane corresponding to a focus. When $\zeta > 1$, the transient rapidly decays to the equilibrium state which is now a node; when $\zeta = 1$ we say that the system has critical damping.

**dangerous bifurcation**

This is a discontinuous bifurcation at which a jump occurs to some remote solution (*see also* **crisis, jump phenomenon**).

*Reference*: Thompson and Stewart (1986).

**degeneracy conditions for 2-dimensional equilibrium**

Let $f : \mathcal{R}^2 \to \mathcal{R}^2$ be a smooth vector field with a non-hyperbolic fixed point at $x^* = 0$. There are four possibilities:

(i) $A(0) = [\partial f_i/\partial x_j]_{i=j=1}^2|_{x=0}$ has real eigenvalues and one of them is zero, i.e. $\det A(0) = 0$, $\mathrm{Tr}A(0) \neq 0$;

(ii) $A(0)$ has pure imaginary eigenvalues, i.e. $\mathrm{Tr}A(0) = 0$, $\det A(0) > 0$;

(iii) both eigenvalues of $A(0)$ are zero but $A(0)$ is not a null matrix, i.e. $\mathrm{Tr}A(0) = \det A(0) = 0$, $A(0) \neq 0$;

(iv) $A(0) = 0$.

The equalities (inequalities) in (i)–(iv) are called degeneracy conditions.

*Example*: Consider a differential equation of normal form:

$$\frac{\mathrm{d}x}{\mathrm{d}t} = x\left(\lambda + \sum_{i=1}^{N-1} a_i y^i\right) + O(|x|^{N+1})$$

and

$$\frac{\mathrm{d}y}{\mathrm{d}t} = \sum_{i=2}^{N} b_i y^i + O(|x|^{N+1})$$

where $N \geq 2$, $a_i$, and $b_i$ are constants, and $b_2 \neq 0$.

A non-hyperbolic fixed point $x^*$ with this normal form is said to be of the saddle–node type. It is characterised by the degeneracy conditions,

$$\det A(x^*) = 0 \text{ and } \mathrm{Tr}A(x^*) \neq 0,$$

and the non-degeneracy condition,

$$b_2 \neq 0.$$

*Reference*: Arrowsmith and Place (1990).

**degenerate Hopf bifurcation** (*see* **Hopf bifurcation**)

**degenerate node**
A node (*see* **fixed points**) with equal real eigenvalues is called a degenerate node. The names improper node and inflected node are also used.

**degrees of freedom**
The number of independent coordinates necessary to describe the position and momentum of the system in the Euclidean space is referred to as the number of degrees of freedom.

*Example*: A particle moving on a line has one degree of freedom. The same particle moving in the three-dimensional Euclidean space has three degrees of freedom. The rigid body in the three-dimensional Euclidean space has six degrees of freedom (three translations and three rotations).

*Remark*: In order to describe the dynamics an $N$-degree-of-freedom system one needs $2N$ first order ordinary differential equations so that the corresponding dynamical system is $2N$-dimensional.

**degree of ODE**
The highest-order derivative in an ordinary differential equation (ODE) indicates the degree of the equation.

*Example*: The degree of the following equation is equal to $n$

$$\frac{\mathrm{d}^n x}{\mathrm{d}t^n} + \frac{\mathrm{d}^{n-1} x}{\mathrm{d}t^{n-1}} + \dots + x = 0.$$

**delay coordinates reconstruction**

An experimental system for which the equations of motion are unknown presents obvious difficulties when we try to mathematically model the dynamics. In this case, the attractor corresponding to a steady state has to be reconstructed from the measured time series $z(t)$. The idea, which is justified by embedding theorems (Takens (1980); Whitney (1936)), is as follows: for a generic observable $z(t)$ and time delay $\tau$, an $m$-dimensional portrait constructed from the vectors

$$[z(t_0), z(t_0 + \tau), ...., z(t_0 + (m-1)\tau)]$$

can have the same properties (the same Lyapunov exponents) as the original attractor. The graphical description of this idea for the *embedding dimension* $m = 3$ is shown in Figure 35.

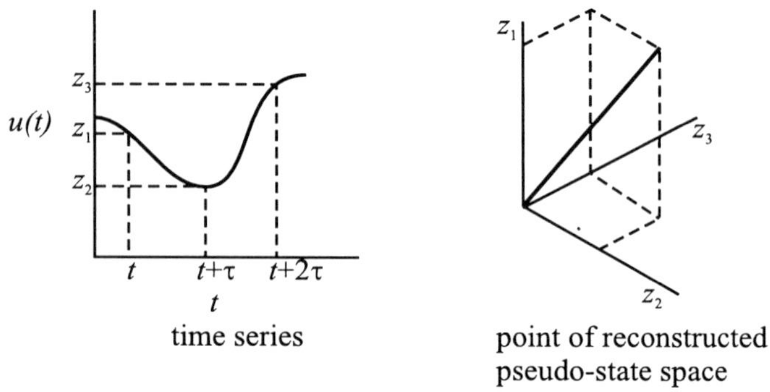


**Figure 35** Reconstruction of the attractor from time series

*Example*: In Figure 36, we consider the projection of the attractor of the Rössler equations for $a = b = 0.2, andc = 5.7$ (*see* **Rössler equations**). In Figure 36(a), we have the attractor obtained from direct integration of the Rössler equations, while Figure 36(b) shows the attractor reconstructed from time series.

*Remark*: Strictly speaking, the phase portrait obtained by this procedure gives an embedding of the original manifold. The choice of the time delay $\tau$ is almost, but not completely, arbitrary.

If we have a system modelled by partial differential equations, or we have experimental time series, we do not know how to choose the embedding dimension $m$. According to embedding theorems, if it is possible to follow $N$ independent variables, then $m$ satisfies the inequality

$$m > 2N + 1.$$

(a)

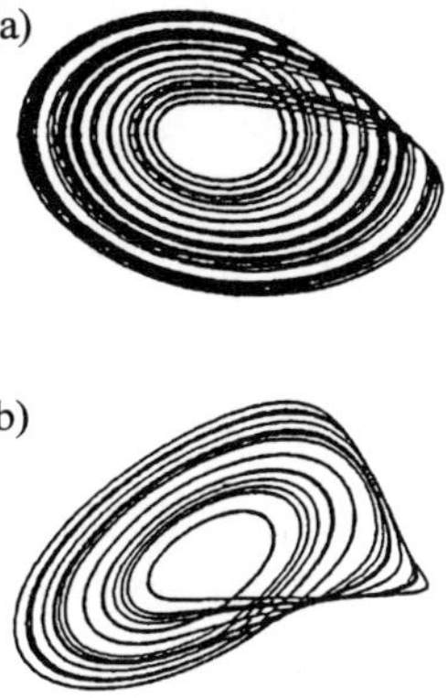

(b)

**Figure 36** Rössler attractors: (a) original; (b) reconstructed from a generated time series

However, in most cases an unambiguous phase portrait can be obtained with fewer dimensions than the inequality requires.

In practice $m$, is increased sequentially until the correlation dimension (*see* **dimension of sets**) fails to increase.

*References*: Takens (1981); Whitney (1936).

**delay-differential equation**

A differential equation of the form

$$\frac{\mathrm{d}x}{\mathrm{d}t} = f(x(t), x(t-\tau))$$

where $x \in \mathcal{R}^n$ and $\tau$ is constant, is called a delay-differential or retarded equation. One has to specify the initial condition for a delay-differential equation $x$ on the interval $[-\tau, 0]$.

The phase space of a delay-differential equation is infinite dimensional.

**Denjoy's theorem**

If an orientation-preserving diffeomorphism of a circle $f : \mathcal{S}^1 \to \mathcal{S}^1$ is of class $C^2$ and the rotation number $\rho(f) = \beta \in \mathcal{R} \setminus \mathcal{Q}$, i.e. is irrational, then $f$ is topologically conjugate to the pure rotation $R_\beta(\theta) = (\theta + \beta) \bmod 1$.

*Reference*: Arrowsmith and Place (1990).

**dense set**

(i) The set $A \subseteq Y$, where $Y$ is a metric space, is a dense subset of $Y$, if, for every $y \in Y$ and every $\delta > 0$, there is some $a \in A$ satisfying $|a - y| < \delta$.

(ii) $A$ is a dense subset of $Y$, where $Y$ is a topological space, if, for any open subset $S \subseteq Y$, the intersection $S \cap A$ is non-empty.

*Remark*: Definitions (i) and (ii) are equivalent.

The set $A \subset \mathcal{R}^n$ is dense in itself if, in every neighbourhood of any point $a \in A$, there is another point of $A$, i.e. every point of $A$ is a limit point.

An open set $A$ is dense if every point in the compliment of $A$ can be approximated arbitrarily close by points in $A$ (as $A$ is dense in $\mathcal{R}^n$), but no point in $A$ can be approximated arbitrarily close by points in the compliment of $A$ (because $A$ is open).

*Example*: Consider a curve on the plane $\mathcal{R}^2$, e.g. $x = y^2$. The compliment of this curve is a dense open set, $X$. If $x_0 \neq y_0^2$, then there are points $(x, y) \in X$ such that, if $|x_0 - x|$ and $|y_0 - y|$ are sufficiently small, then $x \neq y^2$, so that $X$ is open. In addition, if $(x_0, y_0) \in \mathcal{R}^n$, one can find an $(x, y) \in \mathcal{R}^n$ as close to $(x_0, y_0)$ as desired and such that $x \neq y^2$, thus proving that $X$ is dense.

*Reference*: Iyanaga and Kawada (1980).

**derived set**
The set of limit points of $X$ is called the derived set of $X$, and is denoted by $X'$.

**describing function**
This is used particularly in control and electrical systems where initially a harmonic balance method is utilised. The describing function $\Phi$ of a time-invariant nonlinearity $\phi$ is obtained from

$$\Phi(a) = \frac{2\omega}{\pi a} \int_0^{\pi/\omega} \phi(a \sin \omega t) \sin \omega t dt \text{ for } a \neq 0.$$

*Reference*: Iyanaga and Kawada (1980).

**deterministic system**
If the dynamics of a system can be described without the inclusion of any random processes (functions in the probability space) then such a system is said to be deterministic.

**detuning**
This is a term used in vibration studies of forced oscillators where the detuning $\Delta$ is defined as

$$\Delta = 2(\omega_0 - \omega)$$

where $\omega_0$ is the system's natural frequency and $\omega$ the driving frequency ($\omega$ and $\omega_0$ are usually close in value).

*Reference*: Stokes (1959).

**devil's staircase**
A devil's staircase is a set on the plane constructed in the following way. First, the interval $[0, 1]$ is divided into three equal parts and the middle part is stipulated in such a way that the function is equal to 1/2 (see Figure 37). Then we divide the left and right thirds into three equal parts and stipulate that the function equals 1/4 from 1/9 to 2/9 and equals 3/4 from 7/9 to 8/9. Now we have four intervals in which the function is not defined: [0,1/9], [2/9,1/3], [2/3,7/9] and [8/9,1]. We now divide each of these into three equal parts and set the function equal to 1/8, 3/8, 5/8 and 7/8, respectively, on the four middle pieces. By continuing this process, we will obtain a devil's staircase.

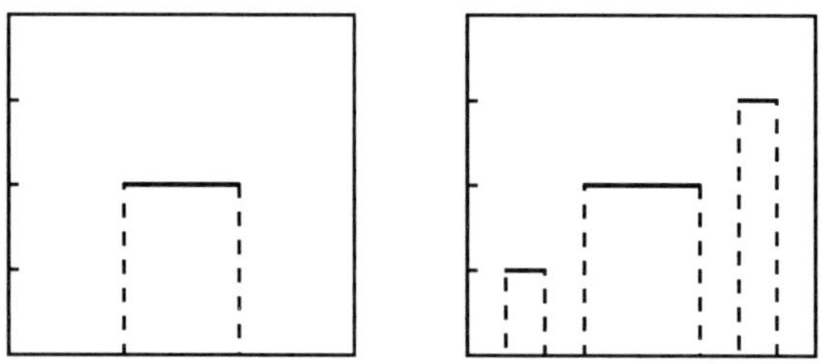

**Figure 37** First steps in the construction of a devil's staircase

*References*: Steeb (1991).

**diffeomorphism**
A $C^k$-diffeomorphism ($k = 0, 1, 2, ...$) $f : M \to N$, is a mapping $f$ which is one-to-one, onto, and has the property that both it and its inverse are $k$ times differentiable, and that the $k$th derivative is continuous.

**difference equation** (*see* **discrete dynamical system**)

**diffusion equation**
The partial differential equation

$$\frac{\partial u(x,t)}{\partial t} = k\frac{\partial^2 u(x,t)}{\partial x^2} - u^3(x,t)$$

where $k$ is constant, is called the nonlinear diffusion equation.

**dimension of sets**
There are several measures which allow estimation of the dimension of sets.

*Topological dimension*

The strict mathematical definition of *topological dimension* is complex. One interpretation is the number of real parameters one has to use in order to indicate the position of a point in a set $A$; this is called a topological dimension $d_T$.

*Remark*: A topological dimension only takes integer values.

*Example*: Let sets $A, B$ and $C$ be, respectively, a point, a curve and a plane. Then $d_T(A) = 0$, $d_T(B) = 1$, and $d_T(C) = 2$.

*Hausdorff dimension*

Let $S$ be a set in $\mathcal{R}^n$. We then define the Hausdorff measure of $S$, indexed by the parameter $\delta \geq \mathcal{R}$, in the following way. Define:

$$l_\delta(S) = \lim_{\epsilon \to 0+} l_{\delta,\epsilon}(S),$$

where

$$l_{\delta,\epsilon}(S) = \inf_{K(\epsilon)} \sum_{B_i \in K(\epsilon)} |B_i|^\delta,$$

in which $|B_i|$ is the volume of the ball $B_i$ and where the infimum bound is taken over all of the coverings $K(\epsilon)$ of the set $S$ made of balls $B_i$ of diameter smaller then $\epsilon$. The *Hausdorff dimension* of $S$, $d_H(S)$, is then defined as the unique value of $\delta$ such that $l_\delta(S)$ is finite, i.e.

$$\delta > d_H(S) \to l_\delta(S) = 0$$

$$\delta < d_H(S) \to l_\delta(S) = +\infty.$$

Let us note that here $d_H$ can take non-integer values. The Hausdorff measure associated with the dimension $d_H$ is $l_{d_H}$, and is associated with the Lebesgue measure in $\mathcal{R}^n$. Thus, in order to evaluate the relative 'sizes' of two given sets, one just needs to compare their Hausdorff dimensions and, if they are equal, the values of their Hausdorff measures.

*Capacity dimension*

The *capacity* or *box-counting dimension* was introduced by Kolmogorov (1958). Let $S$ be a subset of $\mathcal{R}^n$ and $K(\epsilon)$ a covering of $S$ with balls of diameter $\epsilon$. Let $N(\epsilon)$ be the minimal number of balls in $K(\epsilon)$. The capacity of $S$, $d_c(S)$, is then defined as the limit

$$d_c(S) = \lim_{\epsilon \to 0+} \sup \frac{\ln N(\epsilon)}{\ln(1/\epsilon)}.$$

*Example 1*: The Hausdorff dimension and the capacity dimension of the empty set are equal:

$$d_H(\emptyset) = d_c(\emptyset) = -\infty$$

and in this case the topological dimension $d_T$ is not defined.

*Example 2*: For sets such as points, segments of a line or surfaces, the Hausdorff dimension $d_H$ and the capacity dimension $d_c$ are equal to the topological dimension, which is 0, 1 and 2, respectively.

*Example 3*: Consider the triadic Cantor set. At the $n$th step of the construction process, the set is made of $2^n$ intervals of equal length $3^{-n}$. Thus, the Lebesgue measure of the triadic Cantor set obtained when $n \to \infty$ is 0, and its topological dimension $d_T$ is also 0. Let us consider the covering corresponding to the $n$th step of the construction process, which is made of

$$N(\epsilon) = 2^n$$

intervals of size $\epsilon = 3^{-n}$. When $\epsilon$ goes to 0 ($n \to \infty$), one thus deduces that the capacity dimension of the triadic Cantor set is

$$d_c = \frac{\ln 2}{\ln 3} = 0.6309....$$

The Hausdorff dimension can be obtained in a similar way and is found to be the same as $d_c$, so the triadic Cantor set is then characterised by the following dimensions

$$d_T = 0$$

and

$$d_H = d_c = \frac{\ln 2}{\ln 3}.$$

*Information dimension*

The capacity dimension gives the scaling of the number of cubes needed to cover an attractor. In the case of a strange attractor the frequency with which different cubes are visited is vastly different from cube to cube, so for very small $\epsilon$ it is common that only a very small number of the cubes needed to cover the chaotic attractor contain the vast majority of the natural measure on the attractor. In order to take into account different natural measures of the cubes, it is necessary to introduce another definition of dimension, namely the *information dimension.* The quantity

$$d_I = \lim_{\epsilon \to 0} \frac{-\sum_{i=1}^{N(\epsilon)} \mu_i \ln \mu_i}{\ln(1/\epsilon)}$$

where $\mu_i = 1/N(\epsilon)$, is called the information dimension; $\mu_i$ can be considered as a probability with which a trajectory visits the $i$th cube.

*Correlation dimension (Grassberger-Procaccia dimension)*

Let $z_k$ be a trajectory on an attractor. We can compute the correlation integral which is approximated by a sum

$$C(\epsilon) = \lim_{k \to \infty} \frac{1}{K^2} \sum_{i,j}^{K} U(\epsilon - |z_i - z_j|)$$

where $U(.)$ is the unit step function ($U(x) = 1$ for $x = 0$ and $U(x) = 0$ for $x \neq 0$). The quantity $C(\epsilon)$ may be shown to scale with $\epsilon$ in the following way

$$d_{corr} = \lim_{\epsilon \to 0} \frac{\ln(C(\epsilon))}{\ln \epsilon}$$

where $d_{corr}$ is called the *correlation dimension.*

*References*: Edgar (1990); Falconer (1985); Grassberger and Procaccia (1983); Kolmogorov (1958); Mandelbrot (1982).

**discrete dynamical system**
A discrete dynamical system is one in which the evolution of the variables is measured in discrete steps. The behaviour of the system is governed by an equation of the form:

$$x_{n+1} = f(x_n)$$

which is known as a difference equation, an iterated map, or simply a map.

The sequence of points, $x_0, x_1, x_2, ...$ etc., is called an orbit. The term trajectory is also used but usually this refers specifically to orbits of continuous systems, i.e. flows.

**dissipative system** (*see* **gradient system**)

**divergence**

*Sense 1*: If a trajectory moves away from a specific point in phase space then it may be said to diverge, e.g. a trajectory which evolves towards an attractor at infinity.

*Sense 2*: Let $\mathrm{d}x/\mathrm{d}t = f(x)$, where $x \in D \subset \mathcal{R}^n$. The divergence of the vector field $f(x)$ at each point $x \in D$ is given by

$$\nabla \cdot f = \frac{\partial f_1}{\partial x_1} + \frac{\partial f_2}{\partial x_2} + ... + \frac{\partial f_n}{\partial x_n}.$$

In dissipative systems $\nabla \cdot f < 0$.

**double point**

Consider a dynamical system

$$\frac{\mathrm{d}x}{\mathrm{d}t} = f(x, c)$$

where $x \in \mathcal{R}^n$, and $c \in \mathcal{R}$. Let a bifurcation point $(x_0, c_0)$ be a fixed point, $\partial f/\partial x|_0 = 0$ and $\partial f/\partial c|_0 = 0$. In the neighbourhood of the bifurcation point we can set $x = x_0 + \mathrm{d}x$, and $c = c_0 + \mathrm{d}c$ and expand $f(x, c)$. Then the lowest order terms remaining in the Taylor expansion are

$$f(x_0 + \mathrm{d}x, c_0 + \mathrm{d}c) = \frac{1}{2}\left(\frac{\partial^2 f}{\partial x^2}\right)_0 (\mathrm{d}x)^2 + \left(\frac{\partial^2 f}{\partial x \partial c}\right)_0 \mathrm{d}x\mathrm{d}c + \frac{1}{2}\left(\frac{\partial^2 f}{\partial c^2}\right)_0 (\mathrm{d}c)^2 =$$

$$= \alpha(\mathrm{d}x)^2 + 2\beta \mathrm{d}x\mathrm{d}c + \gamma(\mathrm{d}c)^2. \qquad (1)$$

If $(\mathrm{d}x, \mathrm{d}c)$ are along the solution set, so that $f(x, c) = 0$, then (1) yields two distinct roots for the tangents (e.g. $\mathrm{d}c/\mathrm{d}x$) if and only if

$$D = \beta^2 - \alpha\gamma > 0. \qquad (2)$$

If (2) is satisfied, the bifurcation point is called a double point. The bifurcation which occurs at double point is transcritical.

*Reference*: Jackson (1990).

**double-scroll attractor**
Attractor formed by the chaotic response of the Chua or double-scroll circuit – *see* **Chua's circuit**.

**dripping tap**
A common experiment for displaying interesting dynamics can readily be carried out by using a dripping tap (or *leaky faucet* in American parlance). The flow rate is adjusted while the rate at which drops emerge is monitored. Controlled experiments reveal a detailed period-doubling sequence and chaotic behaviour.

*Reference*: Alligood *et al.* (1997).

**driven oscillator**
The dynamical system described by an ordinary differential equation of the form

$$\frac{\mathrm{d}^2 x}{\mathrm{d}t^2} + b\frac{\mathrm{d}x}{\mathrm{d}t} + k(x) = F(t)$$

where $b$ is constant, $x \in \mathcal{R}$ and $F(t)$ is a periodic function, is called a driven oscillator.

**dry friction**
This friction is a friction between two surfaces due to the microscopic interaction of the sliding surfaces (*see* **damping**).

**dual cusp**
One of the elementary catastrophes (*see* **Thom's theorem**).

**Duffing equation**
The second-order differential equation

$$\frac{\mathrm{d}^2 x}{\mathrm{d}t^2} + a\frac{\mathrm{d}x}{\mathrm{d}t} + bx + cx^3 = 0 \tag{1}$$

where $a, b$ and $c$ are constant, and $a \geq 0$, is called the Duffing equation.

The periodically forced Duffing equation has the following form

$$\frac{\mathrm{d}^2 x}{\mathrm{d}t^2} + a\frac{\mathrm{d}x}{\mathrm{d}t} + bx + cx^3 = B\cos\omega t \tag{2}$$

where $B$ and $\omega$ are, respectively, the amplitude and frequency of the forcing.

*Remark*: The Duffing equation describes the dynamics of a number of engineering systems and is sometimes referred to as the Duffing oscillator.

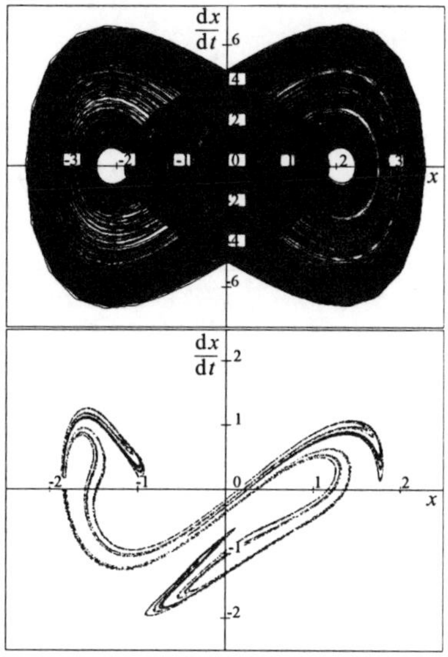


**Figure 38** (a) Phase portrait and (b) Poincaré map of the Duffing equation

*Example*: For $a = 0.1$, $b = 0$, $c = 1$, $B = 10$ and $\omega = 1$, the forced Duffing equation (2) displays chaotic behaviour, as seen in the phase portrait of Figure 38(a) and Poincaré map of Figure 38(b).

*References*: Guckenheimer and Holmes (1983); Kapitaniak (1991).

### Duffing double-well equation

The Duffing equation, written as

$$\frac{\mathrm{d}^2x}{\mathrm{d}t^2} + a\frac{\mathrm{d}x}{\mathrm{d}t} + x - x^3 = 0$$

where $a$ is constant, so that the potential energy function has two symmetric potential wells (as shown in Figure 39), is called the Duffing double-well equation.

### Dulac's criterion

This criterion shows that if the function $B(x, y)$ is $C^1$ and is such that $\partial(BP)/\partial x + \partial(BQ)/\partial y$ does not change sign in a simply connected region, then there is no periodic solution of

$$\frac{\mathrm{d}x}{\mathrm{d}t} = P(x, y)$$

and

$$\frac{\mathrm{d}y}{\mathrm{d}t} = Q(x, y)$$

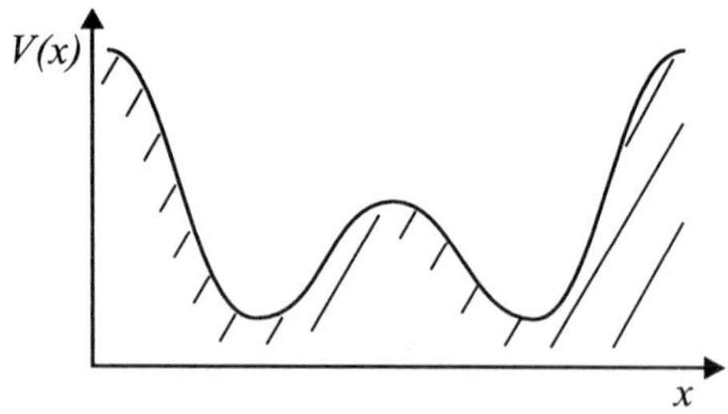


**Figure 39** Potential function of Duffing double well

in that region.

*Reference*: Jackson (1990).

**dynamical system**
The dynamical system defined by

$$\frac{du}{dt} = f(u)$$

where $u \in D \subset \mathcal{R}^n$, is the mapping

$$\Phi : \mathcal{R}^+ \times D \to \mathcal{R}^n$$

which is defined by the solution $u(t) = \Phi(t, u(0))$.

*Example*: Consider the harmonic oscillator

$$\frac{d^2u}{dt^2} + u = 0, \qquad u_0 = u(0), \quad \frac{du(0)}{dt} = \dot{u}_0.$$

In order to obtain the corresponding system of differential equations, we put $u = u_1$, and $du/dt = u_2$. This yields

$$\frac{du_1}{dt} = u_2, \qquad \frac{du_2}{dt} = -u_1.$$

The solution of the initial-value problem is given by

$$u_1(t) = u_0 \cos t + \dot{u}_0 \sin t$$

and

$$u_2(t) = -u_0 \sin t + \dot{u}_0 \cos t.$$

Thus, the dynamical system is characterised by the mapping

$$\Phi(t, u_0, \dot{u}_0) = (u_0 \cos t + \dot{u}_0 \sin t, -u_0 \sin t + \dot{u}_0 \cos t)$$

where $\Phi : \mathcal{R}^+ \times \mathcal{R}^2 \to \mathcal{R}^2$.

**dynamical systems theory**
This theory is the collection of ideas, theorems and numerical algorithms which are applied to dynamical systems.

# E

## ECG

Electrocardiography: the measurement of electrical activities of the heart which has been shown to display periodic and chaotic behaviour (see Figure 40).

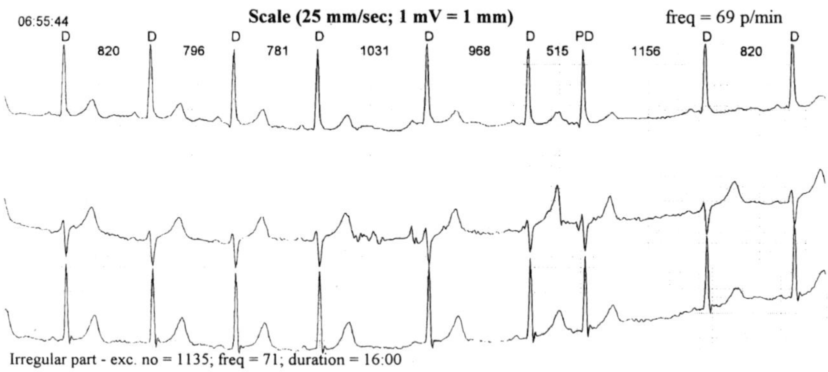


**Figure 40** Irregular part of the ECG record of a 72 year old man

*Reference*: Glass and Mackey (1988).

## EEG

Electroencephalography: the measurement of electrical activity in the brain which has been shown to display temporal and spatial patterns (see Figure 41).

*Reference*: Glass and Mackey (1988).

## eigenvalue and eigenvector

Let $A$ be an $n \times n$ dimensional matrix. The eigenvalues $\lambda_1 \ldots, \lambda_n$ are the solutions of the characteristic equation

$$\det(A - \lambda I) = 0, \tag{1}$$

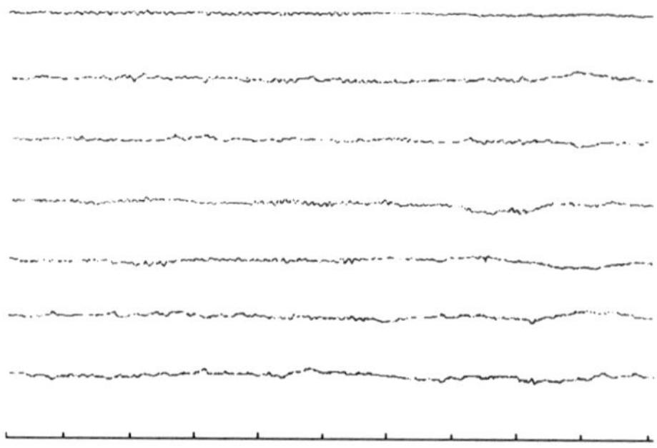

**Figure 41** Example of an EEG record

where $I$ is the $n \times n$ identity matrix. Any vector $\mathbf{v} \neq 0$ satisfying the equation

$$(A - \lambda I)\mathbf{v} = 0 \tag{2}$$

is called an eigenvector.

*Example*: Consider the equation $\mathrm{d}u/\mathrm{d}t = Au$, where

$$A = \begin{pmatrix} 1 & 1 \\ 4 & 1 \end{pmatrix}$$

In order to find the eigenvalues, we have to consider the matrix

$$A - \lambda I = \begin{pmatrix} 1-\lambda & 1 \\ 4 & 1-\lambda \end{pmatrix}$$

and obtain

$$\det(A - \lambda I) = (1-\lambda)^2 - 4 = 0.$$

Thus, the eigenvalues are $\lambda_1 = -1$ and $\lambda_2 = 3$. Let $v_1$ be the corresponding eigenvector for $\lambda_1 = -1$. Then (2) becomes

$$\begin{pmatrix} 2 & 1 \\ 4 & 2 \end{pmatrix} \mathbf{v}_1 = 0.$$

A non-trivial solution is

$$\mathbf{v}_1 = \begin{pmatrix} 1 \\ -2 \end{pmatrix}.$$

Note that there is no unique eigenvector corresponding to $\lambda_1 = -1$, since any multiple of $\mathbf{v}_1$ is also an eigenvector. In the similar way, one can also find the eigenvector for $\lambda_2 = 3$.

*Reference*: Hirsch and Smale (1974).

**elastica**
This is a long thin flexible beam displaying in-plane and out-of-plane motion when compressed. Spatial behaviour of beams is a fundamental problem in civil engineering. The configuration of the beam, assumed to be planar, is most conveniently described by $u(\xi)$, the angle which the beam makes with the horizontal, as a function of the arc length $\xi$ (*see* Figure 42).

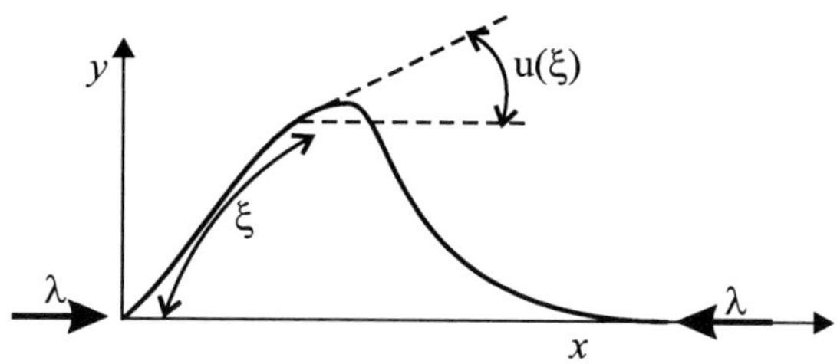


**Figure 42** Coordinates on the beam

Let us normalise the rod to have length $\pi$. The displacement $(x(\xi), y(\xi))$ may be calculated from the formulae

$$x(\xi) = \int_0^{\xi} \cos u(\xi')\mathrm{d}\xi', \qquad y(\xi) = \int_0^{\xi} \sin u(\xi')\mathrm{d}\xi'.$$

The equilibria of the beam are characterised by the two-point boundary problem

$$-\frac{\mathrm{d}^2 u}{\mathrm{d}\xi^2} - \lambda \sin u = 0 \quad u'(0) = u'(\pi) = 0 \tag{1}$$

where $\lambda$ is the compressive force applied to the beam. This equation is just the first variation of a minimisation problem with constraints. It is derived under the following two assumptions:

(i) The beam is incompressible but capable of bending, with the stored energy function being proportional to

$$\int_0^{\pi} \kappa^2(\xi)\mathrm{d}\xi$$

where $\kappa = \mathrm{d}u/\mathrm{d}\xi$ is the curvature.

(ii) The ends of the rod are hinged free, thus freely permitting rotation, but are constrained to lie on a fixed line.

The last equation is mathematically identical to that of a planar mathematical pendulum. There is, however, one important difference between the two equations because the pendulum is an initial-value problem while the elastica is strictly speaking a boundary-value problem. Under the assumption that our elastica is infinite, we can treat (1) as an initial-value problem:

$$\frac{\mathrm{d}^2u}{\mathrm{d}\xi^2} + \sin u = 0. \tag{2}$$

The spatial plots representing the above are shown in Figure 43.

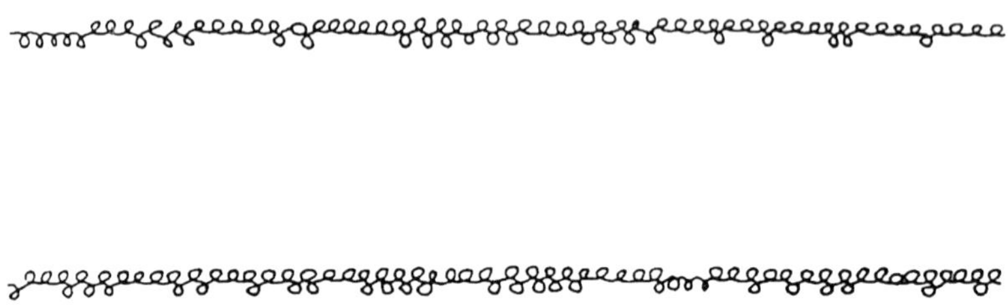

**Figure 43** The spatial plots for equation (2)

*References*: El Naschie (1988); El Naschie and Al Athel (1989).

**elliptic functions** (*see* **Jacobi elliptic functions**)

**elliptic integrals**
These integrals arise in the analytical solutions of ODEs. The elliptic integral of the first kind is

$$F(x,k) = \int_0^x \frac{\mathrm{d}y}{[(1-y^2)(1-k^2y^2)]^{1/2}}$$

or

$$\bar{F}(\phi,k) = \int_0^\phi \frac{\mathrm{d}\theta}{(1-k^2\sin^2\theta)^{1/2}}$$

where $k^2 < 1$, and the transformation $x = \sin\phi$ connects the two variables.

The elliptic integral of the second kind is

$$E(x,k) = \int_0^x \left(\frac{1-k^2y^2}{1-y^2}\right)^{1/2} \mathrm{d}y$$

or

$$\bar{E}(\phi,k) = \int_0^\phi (1-k^2\sin^2\theta)^{1/2}\mathrm{d}\theta.$$

The quantity $k$ ($k^2 < 1$) is called the modulus of the elliptic integral of the third kind and $k'$, given by $k'^2 = 1 - k^2$, the complementary modulus.

*Reference*: Byrd and Friedman (1971).

**elliptic point**

This is a different name for centre (*see* **fixed points**).

**El-Niño event**

El-Niño may be defined as the appearance of anomalously warm water in the eastern equatorial Pacific. Associated with this is a weakening and sometimes a reversal of the trade-wind field. Major El-Niño-southern-oscillation (ENSO) have events occurred in 1957, 1965, 1972 and 1982.

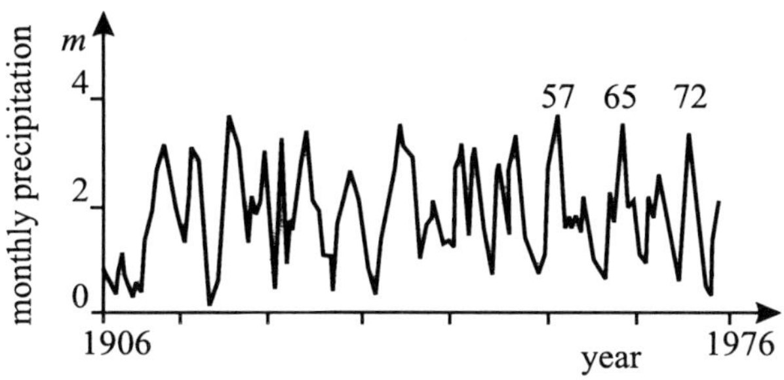


**Figure 44** Mean monthly precipitation on Nauru Island

The various events differ in detail and intensity but appear to have broadly similar overall features. El-Niño has major economic consequences and possibly global climatic effects. This type of event can be determined by the observation of certain meteorological data. For example, Figure 44 demonstrates the variability of rainfall on Nauru Island in the West Pacific. The time series shows the characteristic that usually very little, if any, precipitation falls in this region of the central Pacific, with large amounts of precipitation only occurring during El-Niño.

A simple, but realistic model can be described which explains all of the broad qualitative features of the phenomenon. Imagine an equatorial ocean to be a box of fluid characterised by temperatures in the east and west ($T_e$ and $T_w$) and a current $u$, as shown in Figure 45. The current is driven by a surface wind $U$, which is in part generated by the temperature gradient $(T_e - T_w)\Delta x$, where $\Delta x$ is the distance between points $e$ and $w$. A cooler temperature in the east ($T_e < T_w$) produces a westward surface wind across the ocean, because

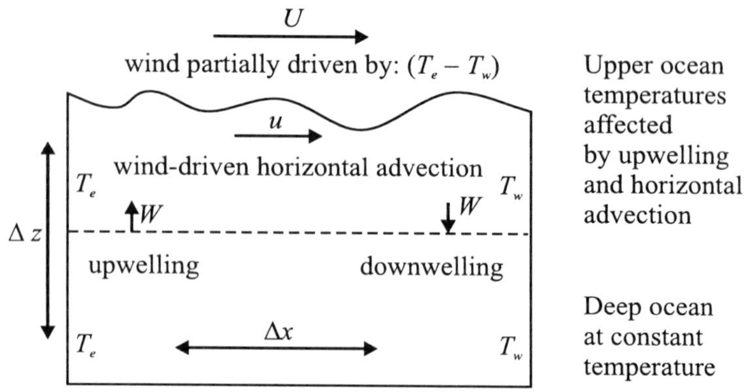


**Figure 45** Model of the ocean-atmosphere system

of the convective tendency for air to rise (sink) over warm (cool) water. Thus we write:

$$\frac{\mathrm{d}u}{\mathrm{d}t} = \frac{B(T_e - T_w)}{2\Delta x} - C(u - u^*)$$

where $B$ and $C$ are constants. The term

$$[B(T_e - T_w)/\Delta x] + Cu^*$$

represents wind-produced stress, and $-Cu$ represents mechanical damping, while a negative value for the constant $u$ would represent the effect of the mean tropical surface easterly. Variations in pressure have been neglected as they do not qualitatively affect the model behaviour.

The temperature field is advected by the current. Assuming a deep ocean of constant temperature $T$, the simplest finite difference approximation to the temperature equation of fluid flow is

$$\frac{\mathrm{d}T_w}{\mathrm{d}t} = \frac{u(T - T_e)}{2\Delta x} - A(T_w - T^*)$$

and

$$\frac{\mathrm{d}T_e}{\mathrm{d}t} = \frac{u(T_w - T)}{2\Delta x} - A(T_e - T^*).$$

The first terms on each right-hand side represent horizontal advection and upwelling, and the second terms forcing and thermal damping ($A$ and $T^*$ are constants). $T^*$ is the temperature to which the ocean would relax in the absence of motion and therefore represents radiative processes and heat exchange with the atmosphere.

*Reference*: Vallis (1988).

**embedding**
A homeomorphism $f$ which maps the set $X$ into a space $Y$ is called a homeomorphic embedding of $X$ and $Y$.

If $X$ has dimension $\dim X$ and $\dim X \leq n$, then $X$ can be embedded in $Y$ if $\dim Y \leq 2n+1$. Furthermore, the homeomorphisms are dense in the sense that every mapping of $X$ into $Y$ can be made a homeomorphism by an arbitrary small modification.

A differentiable map $f : X \to Y$ of a smooth manifold $X$ into $Y$ is called an embedding of $X$ in $Y$ if $f(X) \subset Y$ is a differentiable submanifold of $Y$ and $f : X \to f(X)$ is a diffeomorphism.

An $m$-dimensional differentiable manifold can be embedded in $\mathcal{R}^n$ if $n \geq 2m$.

*References*: Jackson (1990); Whitney (1936).

**endomorphism**
The word endomorphism is used for measure preserving, but not necessarily invertible maps.

*Reference*: Lasota and Mackey (1985).

**entrainment**
A different name for mode locking (*see* **mode locking**).

**entropy** (*see* **measure theoretic entropy, topological entropy**)

**equilibrium**
Consider the dynamical system

$$\frac{\mathrm{d}x}{\mathrm{d}t} = f(x)$$

where $x \in \mathcal{R}^n$. A point $x^*$ such that

$$f(x^*) = 0$$

is called an equilibrium point (*see also* **fixed points**).

**ergodic theory**
This theory investigates the statistical properties of a measure-preserving mapping $T^t$ of the phase space $M$ on to itself.

*Example*: The Hamiltonian equations of motion

$$\frac{dq_j}{dt} = \frac{\partial H}{\partial p_j}$$

and

$$\frac{dp_j}{dt} = -\frac{\partial H}{\partial q_j}$$

with $j = 1, ..., N$, gives a mapping $T^t$ of the phase space $M$ on itself, since for each point, $(p, q) = x \in M$ determines an appropriate point $x_t = T^t x \in M$, where $T_x M$ is a manifold tangent to $M$ in the point $x$, and due to the Liouville theorem the initial phase volume is preserved.

A system is ergodic under the map $f$ if the measure of every invariant set, $A = f(A)$ mod 0, is either $\mu(A) = 1$ or $\mu(A) = 0$.

*Remark*: The above definition says that for ergodic systems invariant sets either have measure zero, or else represent all states except for a set of measure zero.

An endomorphism $T$ is ergodic if every measurable set $A$ which is invariant under $T$ $(T^{-1}A = A)$ has either measure zero or 1.

*References*: Jackson (1990); Steeb (1991).

**escape to infinity**
Consider the dynamical system

$$\frac{dx}{dt} = f(x)$$

where $x \in \mathcal{R}^n$. The situation in which the unbounded solution $x(t)$ grows to infinity as $t \to \infty$ is referred to as escape to infinity.

**escape equation**
The second-order differential equation

$$\frac{d^2x}{dt^2} + \frac{dx}{dt} + x - x^2 = F \sin \omega t$$

is referred to as the escape equation.

This equation, which is originaly attributed to Helmholz and recently the focus of intensive study by Thompson and co-workers, is sometimes also called referred Thompson–Helmholz equation, describes the escape of a particle from a canonical potential well.

*Reference*: Thompson (1989).

**Euler method**
This method is the simplest method for numerical integration of ordinary differential equations $\mathrm{d}x\mathrm{d}t = f(x,t)$, given $x_0 = x(t = 0)$. The numerical solution is given by the iterates

$$t_n = t_0 + nh$$

and

$$x_{n+1} = x_n + hf(x_n, t_n)$$

where $n = 1, 2, \ldots$, and the integration step $h$ is constant.

*Reference*: Press *et al.* (1986).

**excitability**
The ability of a system to respond to a sufficient stimulus producing some large excursion and then return to its prior, relatively quiescent state. There are many degrees and kinds of excitability.

*Reference*: Winfree (1988).

**excitable medium**
An excitable system distributed in space with coupling between adjacent cells/pieces so that excitation can propagate as a wave.

*Reference*: Winfree (1988).

**existence and uniqueness theorem for solutions to ODEs**
Let $U \subset \mathcal{R}^n$ be an open subset of real Euclidean space (or of a differentiable manifold $M$). Let $f : U \to \mathcal{R}^n$ be a continuously differentiable $C^1$ map and let $x_0 \in U$. Then there exists a constant $c > 0$ and a unique solution $\phi(x_0, .) : (-c, c) \to U$ satisfying the differential equation

$$\frac{\mathrm{d}x}{\mathrm{d}t} = f(x)$$

with the initial condition $x(0) = x_0$.

*Reference*: Hirsch and Smale (1974).

**explosive instabilities**
This is a different name for crisis (*see* **crisis**).

**exponential divergence**
This is the property of chaotic systems where two initially close trajectories rapidly (exponentially) diverge and become uncorrelated (*see* **sensitive dependence on initial conditions**).

**extended phase space**
Consider the dynamical system

$$\frac{d^2x}{dt^2} + f(\frac{dx}{dt}, x, t) = 0.$$

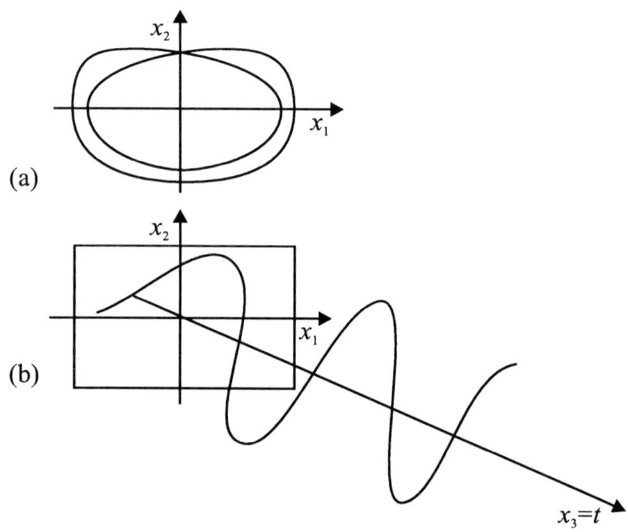


**Figure 46** (a) Phase plane and (b) extended phase space

The phase-space trajectory can be visualised in the phase plane defined by the $(x_1, x_2)$ plane where $x_1 = x$ and $x_2 = dx/dt$. If we additionally put $x_3 = t$, we will have the three-dimensional dynamical system

$$\frac{dx_1}{dt} = x_2$$

$$\frac{dx_2}{dt} = -f(x_1, x_2, x_3)$$

and

$$\frac{dx_3}{dt} = 1.$$

The space $(x_1, x_2, x_3)$ is called the /extended phase space or full phase space. The difference between phase plane and extended phase space is illustrated in Figure 46.

# F

**Farey sequence**
Let $n$ be a natural number and set $x_0 = 0$, $y_0 = x_1 = 1$, and $y_1 = n$; then produce further iterates according to the formulae

$$x_{k+2} = \left[ \frac{y_k + n}{y_{k+1}} \right] x_{k+1} - x_k$$

$$y_{k+2} = \left[ \frac{y_k + n}{y_{k+1}} \right] y_{k+1} - y_k$$

where $[a]$ denotes the greatest integer not greater than $a$. The quantity $x_j/y_j$ is called the Farey fraction, and $x_0/y_0$, $x_1/y_1$,... etc. is called the Farey sequence.

*Example*: If $n = 5$, then in the interval [0,1] we have the following Farey sequence: 0/1, 1/5, 1/4, 1/3, 2/5, 1/2, 3/5, 2/3, 3/5, 2/3, 3/4, 4/5, 1/1.

*Reference*: Falconer (1985).

**fast variables** (*see* **averaging methods**)

**fat fractals** (*see* **fractal sets**)

**faucet** (*see* **dripping tap**)

**feedback controlling methods** (*see* **controlling chaos**)

**Feigenbaum's constant**
A sequence of period-doubling bifurcations leading to chaos has an interesting property. Consider the following ratio:

$$\rho_n = \frac{a_n - a_{n-1}}{a_{n+1} - a_n}$$

where $a_n$ are the values of the parameter at which the period-doubling bifurcations occur, as shown in Figure 47. It was observed by Feigenbaum that

$$\lim_{n\to\infty} \rho_n = \delta$$

where $\delta = 4.66920\ldots$ . Feigenbaum's numerical computations and the analytical results of Collet, Eckmann and Lanford (1970) based on renormalisation theory (*see* **renormalisation group theory**) show that the period-doubling cascade has some universal properties.

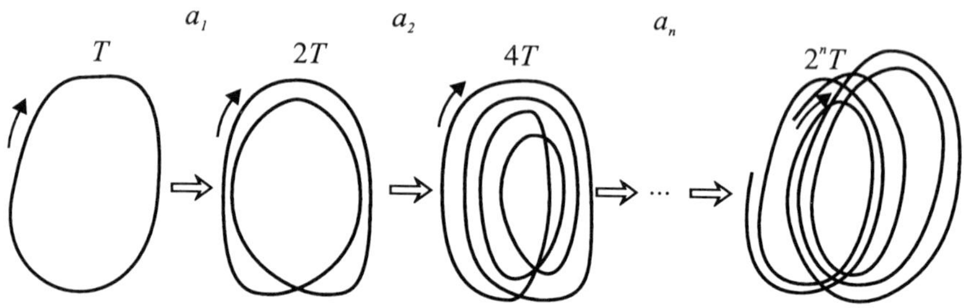


**Figure 47** Cascade of period-doubling bifurcations as a parameter value is varied

*References*: Collet *et al.* (1970); Feigenbaum (1978).

**Feigenbaum's scenario**

The cascade of period-doubling bifurcations viewed in the form of a bifurcation diagram, i.e. plotting a representative state variable $x$ versus a bifurcation parameter $a_i$, as shown in Figure 48, is called Feigenbaum's scenario (*see also* **Feigenbaum's constant**).

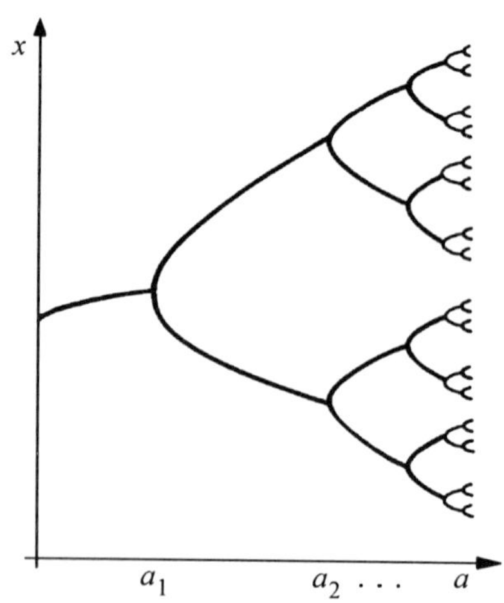


**Figure 48** The Feigenbaum's scenario

**Fermi–Pasta–Ulam model**

The system with the Hamiltonian function

$$H(p,q) = \sum_{j=1}^{N} \left[ \frac{1}{2} p_j^2 + \frac{1}{2}(q_j - q_{j+1}) + \frac{1}{3}(q_{j+1} - q_j)^3 \right]$$

where $q_j = q_{j+N}$ is a periodic boundary condition and $N = 64$, is called the Fermi–Pasta–Ulam model.

*Reference*: Fermi (1965).

**Fibonacci numbers**

These numbers are defined by the difference equation

$$F_n = F_{n-1} + F_{n-2}$$

where $n \geq 2$, $F_0 = 0$ and $F_1 = 1$.

*Example*: The first Fibonacci numbers are 1, 2, 3, 5.

*Reference*: Schroeder (1984).

**final-state sensitivity**

If a system possesses more than one attracting solution and the boundaries which separate their basins of attraction are fractal then the system is said to possess final-state sensitivity. Two close-by initial conditions chosen close to or on this boundary can arbitrarily evolve to give different solutions.

*Reference*: Ott (1992).

**first integral**

Let $V$ be a vector field on a differentiable manifold, and $g : M \to \mathcal{R}$ be a $C^r$ function ($r \geq 1$). A first integral of $V$ is a $C^1$ function $g : M \to \mathcal{R}$, which is constant along trajectories of the dynamical system defined by $V$, i.e.

$$\frac{\mathrm{d}}{\mathrm{d}t} g(x) = \nabla g \cdot \frac{\mathrm{d}x}{\mathrm{d}t} = \nabla g \cdot V(x).$$

*Example*: Consider the autonomous system

$$\frac{\mathrm{d}x_1}{\mathrm{d}t} = x_2 x_3$$

$$\frac{\mathrm{d}x_2}{\mathrm{d}t} = x_3 x_1$$

and

$$\frac{dx_3}{dt} = x_1 x_2$$

defined on $\mathcal{R}^3$. The associated vector field is given by

$$V = x_2 x_3 \frac{\partial}{\partial x_1} + x_3 x_1 \frac{\partial}{\partial x_2} + x_1 x_2 \partial \partial x_3.$$

The first integrals are

$$g_1(x) = x_1^2 - x_2^2$$

and

$$g_2(x) = x_2^2 - x_3^2.$$

*Reference*: Steeb (1991).

**first-order ODE**
If the highest-order derivative in the ordinary differential equation is the first derivative, then the equation is first order.

*Example*: The differential equation

$$\frac{du}{dt} + u = 0$$

where $u \in \mathcal{R}$, is the first-order ODE.

**first return map** (*see* **Poincaré map, return map**)

**fixed points**
The point $u^* \in \mathcal{R}^n$, such that

$$f(u^*) = 0$$

is called a fixed, equilibrium or critical point of the dynamical system given by $du/dt = f(u)$.

*Example*: Consider a two-dimensional linear system, $du/dt = Au$, where the matrix $A$ is $2 \times 2$ dimensional. As the dimension of the phase space is two, the eigenvalues $\lambda_1$ and $\lambda_2$ are both real or they are complex conjugates. The behaviour of the solutions

$$\mathbf{z}(t) = \begin{pmatrix} c_1 e^{\lambda_1 t} \\ c_2 e^{\lambda_2 t} \end{pmatrix}$$

for $\lambda_1 \neq \lambda_2$, and

$$\mathbf{z}(t) = \begin{pmatrix} c_1 e^{\lambda t} + c_2 t e^{\lambda t} \\ c_2 e^{\lambda t} \end{pmatrix}$$

for $\lambda_1 = \lambda_2 = \lambda$ are very different for various choices of $\lambda_1$ and $\lambda_2$.

If the eigenvalues are real and have the same sign, the fixed point is called a *node.* Moreover, when $\lambda_1 < 0, \lambda_2 < 0$, then the critical point is an attractor i.e. a stable fixed point, and if $\lambda_1 > 0$, $\lambda_2 > 0$, then it is a repellor (see Figure 49). It is easy to verify that in the phase-space the orbits are straight lines through the origin when $\lambda_1 = \lambda_2 = \lambda$.

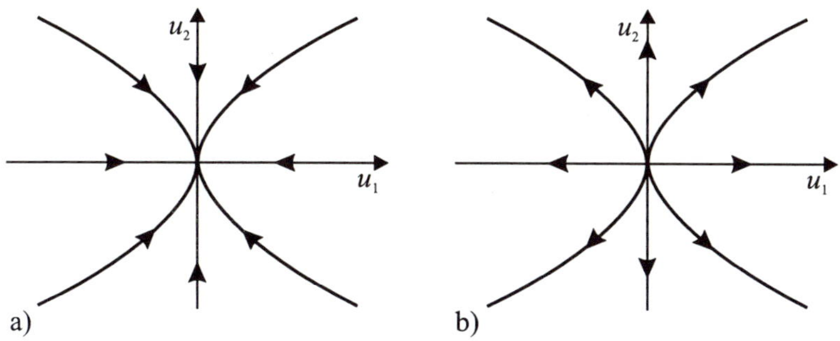


**Figure 49** The node; as: (a) an attractor; (b) a repellor

A fixed point with real, equal eigenvalues is called an improper, degenerate or inflected node.

If the eigenvalues are real but have different sign, then the critical point is called a saddle. In the phase space the orbits are given by

$$|u_1| = c|u_2|^{-|\lambda_1/\lambda_2|}$$

with $c$ as constant. In this case, the critical point is neither an attractor nor a repellor. There exist two solutions with the property $(u_1(t), u_2(t)) \to (0, 0)$ for $t \to \infty$, and two solutions with this property for $t \to -\infty$ (Figure 50). The first two of these solutions are the stable subspace of the *saddle point*, with the other two being the unstable subspace.

When the eigenvalues $\lambda_1$ and $\lambda_2$ are the complex conjugate,

$$\lambda_{1,2} = \mu \pm \omega i$$

with $\mu\omega \neq 0$, the complex solutions are of the form $\exp((\mu \pm \omega i)t)$. The linear combination of the complex solutions leads to real independent solutions of the form

$$e^{\mu t} \cos \omega t, \quad e^{\mu t} \sin \omega t.$$

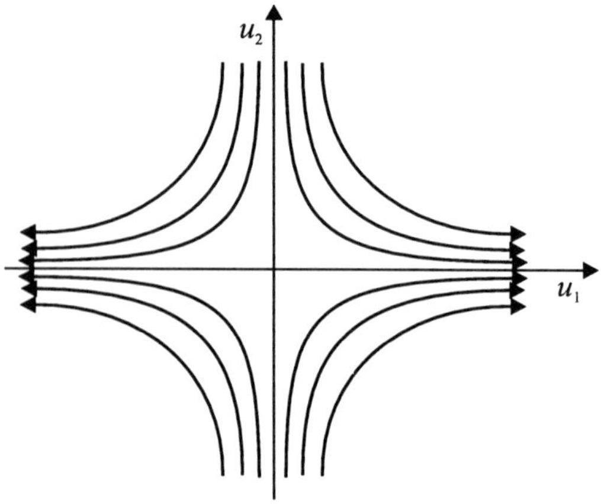


**Figure 50** Saddle point

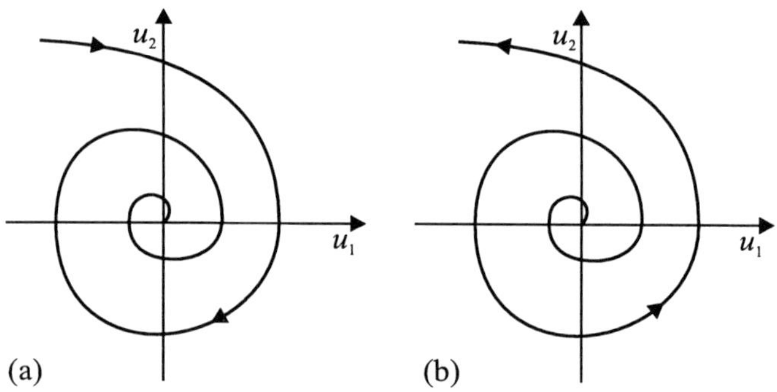


**Figure 51** The focus as: (a) an attractor; (b) a repellor

The orbits spiral in for $\mu < 0$, or out for $\mu > 0$, with respect to the origin, and the critical point is called a *focus* (Figure 51).

The latter case is when the eigenvalues are purely imaginary. If

$$\lambda_{1,2} = \pm \omega i$$

then the critical point is called a *centre*. The solutions can be written as a combination of $\cos \omega t$ and $\sin \omega t$ and the orbits in the phase space are concentric ellipses (Figure 52). It is clear that in this case a critical point is neither an attractor nor a repellor.

*Reference*: Thompson and Stewart (1986).

**flip bifurcation** (*see* **period-doubling bifurcation**)

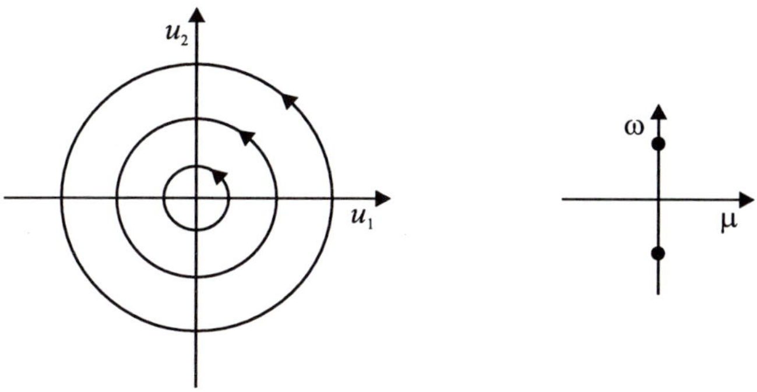


**Figure 52** Representation of the critical point as a centre

**Floquet theory**
This theory is concerned with the solution of ordinary differential equations with periodic coefficients, e.g.

$$\frac{\mathrm{d}^2 y}{\mathrm{d}t^2} + \omega(t)y = 0$$

where $\omega(t+T) = \omega(t)$. If $y_i(t)$ $i = 1, 2$ are two independent solutions, then $y_i(t+T)$ are also independent solutions. Therefore

$$y_i(t+T) = \sum_{j=1}^{2} a_{ij} y_j(t)$$

for the constants $a_{i,j}$. For the general solution

$$F(t) = \sum_{i=1}^{2} c_i y_i(t)$$

with constants $c_i$, one has

$$F(t+T) = \sum_{i,j=1}^{2} a_{ij} c_i y_j(t).$$

The relationship

$$F(t+T) = \sigma F(t)$$

is implied by $\det(A - \sigma I) = 0$, where $A = [a_{ij}]$ and $I$ is the identity matrix. In this situation, bt defining the characteristic Floquet multipliers $\mu$ by $\sigma = \exp(\mu T)$, one can write the general solution as

$$F(t) = \exp(\mu t)\phi(t)$$

with $\phi(t)$ as a $T$-periodic function.

*Remark*: The Floquet theory is sometimes called the Bloch theory.

*References*: Steeb (1991); Jordan and Smith (1988).

**flow**
Let $M$ be a manifold and $\{g^t\}$ be a family of the maps $g^t : M \to M$, such that for all $s, t \in \mathcal{R}$

$$g^{t+s} = g^t g^s$$

and $g^0$ is the identity. The pair $(M, \{g^t\})$ is called a flow.

Consider a dynamical system

$$\frac{\mathrm{d}x}{\mathrm{d}t} = f(x) \tag{1}$$

where $x \in \mathcal{R}^n$ and $f : U \to \mathcal{R}^n$ is a smooth function defined on some subset $U \subseteq \mathcal{R}^n$. We say that the vector field $f$ generates a flow $\phi_t : U \to \mathcal{R}^n$, where $\phi_t(x) = \phi(x, t)$ is a smooth function defined for all $x \in U$ and $t$ in some interval $I = (a, b) \subseteq \mathcal{R}$, and $\phi$ satisfies (1) in the sense that

$$\frac{\mathrm{d}}{\mathrm{d}t}(\phi(x,t))|_{t=\tau} = f(\phi(x,\tau))$$

for all $x \in U$ and $\tau \in I$.

*References*: Arnold (1983); Guckenheimer and Holmes (1983).

**flutter**
Structural self-excited oscillations with growing amplitudes caused by fluids flowing across plates and strings, inside tubes, or involving more complicated structures, are known as flutter.

The occurrence of flutter is often connected with the Hopf bifurcation in various models.

**focus** (*see* **fixed points**)

**fold**
One of the elementary catastrophes (*see* **Thom's theorem**).

The term fold is sometimes used as an abbreviation for a fold bifurcation or saddle–node bifurcation (*see* **bifurcation**).

**folded-band attractor**
The name sometimes used to describe the attractor produced by the Rössler equations (*see* **Rössler equations**).

**follower force**
Consider the clamped-free column in Figure 53, which is loaded by a special type of force which acts along tangents to the free tip of the deformed column. This type of loading is called a follower force.

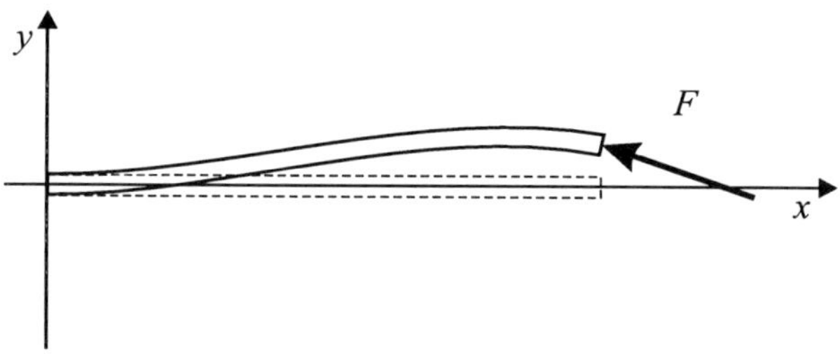


**Figure 53** Column loaded by a follower force

*Reference*: Thomsen (1997).

**forced systems** (*see* **nonautonomous system**)

**fractal dimension**
If the dimension (either Hausdorff or capacity) of the set $A$ is non-integer, it is said to be fractal (*see* **dimension of sets**).

*Reference*: Mandelbrot (1982).

**fractal sets**
Mandelbrot originally suggested that a fractal set should be defined as a set whose Hausdorff dimension is strictly greater than its topological dimension. This definition is adequate for many sets, but there exists a class of sets (of the same type as the triadic Cantor set) whose Lebesgue measure is finite and whose Hausdorff dimension, $d_H = 1$, is equal to the topological dimension, so a new definition had therefore to be introduced.

*Example 1*: Consider the rational numbers in the interval $[0, 1]$. These numbers are dense in $[0, 1]$, since any irrational number can be approximated by a rational number to arbitrary accuracy. The rationals are also countable, since we can arrange them in a linear ordering such as

$$\frac{1}{2}, \frac{1}{3}, \frac{2}{3}, \frac{1}{4}, \frac{3}{4}, \frac{1}{5}, \frac{2}{5}, \frac{3}{5}, \frac{4}{5}, \frac{1}{6}, \frac{1}{7}, \ldots, \text{etc.}$$

To the $n$th rational on this list (denoted by $s_n$), we now associate an interval

$$I_n = \left(s_n - \frac{\eta}{2}\left(\frac{1}{2}\right)^n, s_n + \frac{\eta}{2}\left(\frac{1}{2}\right)^n\right)$$

of length $2^{-n}\,\eta$. We are interested in the set $S$ formed by taking the interval $[0, 1]$ and then successively removing $I_1, I_2, \ldots, I_n$ in the limit $n \to \infty$.

Since the total length of all of the removed intervals is given by

$$\sum_{n=1}^{\infty}\left(\frac{1}{2}\right)^n \eta = \eta$$

we have the Lebesgue measure of $S$, denoted as $\mu(S)$, satisfying

$$\mu(S) > 1 - \eta$$

which is positive if $\eta < 1$. The 'greater than' symbol appears because some of the removed intervals overlap.

As the Lebesgue measure of $S$ is positive, its capacity dimension $d_c$ is the same as the space in which $S$ lies, i.e. $d_c = 1$ and is equal to the Hausdorff and topological dimensions.

Sets similar to that introduced in the above example are called *fat fractals* and can be defined in the following way: a set $S$ lying in an $n$-dimensional Euclidean space is a fat fractal if, for every point $x \in S$, and every $\epsilon > 0$, a ball of radius $\epsilon$ centred at the point $x$ contains a non-zero volume of points in the set and a non-zero volume outside the set.

As fat fractals are not strictly fractals according to the above definition, we say that a fractal set is a set which has some self-similar properties in the sense that its structure is the same on any scale.

This definition is much wider than the one based on the Hausdorff dimension and, as we will see elsewhere, can be easily applied to strange attractors.

*Example 2*: In order to show that chaotic attractors have a fractal structure, let us consider a three-dimensional chaotic attractor $A$ on which a typical trajectory is characterised by the following Lyapunov exponents: $\lambda_1 > 0$,

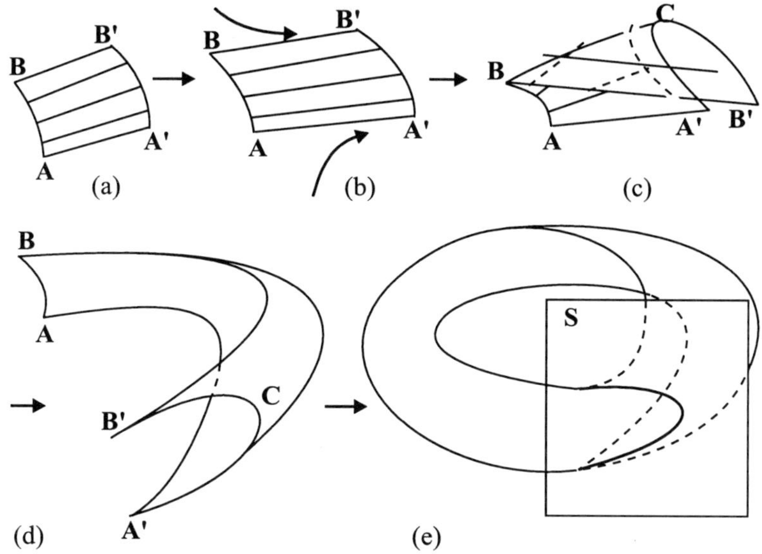


**Figure 54** Geometrical structure of a chaotic attractor

$\lambda_2 = 0$, $\lambda_3 < 0$, and

$$\sum_{i=1}^{3} \lambda_i < 0.$$

Consider the stretching and folding model of the chaotic attractor (*see* **attractor**). We will analyse the evolution of an interval $AB$ in which almost all points belong to the attractor $A$ and evaluate its evolution in time (schematically shown in Figure 54). For small values of $t$, we practically observe no changes in any geometrical structure, with points $A$ and $B$ being transposed, respectively, into $A'$ and $B'$ (Figure 54(b)). As the sum of all Lyapunov exponents is negative the length of $A'B'$ cannot increase to infinity, so we have to observe folding of the surface $ABA'B'$ (Figure 54(c,d)) and the creation of an arc $A'CB'$. The geometrical structure of an attractor can be achieved by connecting points $A'$ and $B'$ (by 'pressing') and 'sticking' point $A'(= B')$ with $A$, and $C$ with $B$. As a result, we have the structure shown in Figure 55(e). Due to the continuity of trajectories on the strange attractor, it is necessary for an infinite number of surfaces like the one shown in Figure 55(e) to exist in the real attractor. In order to detect these surfaces, consider the Poincaré map obtained by the cross-section $S$ shown in Figure 55.

By enlarging parts $S'$ and $S'$, we can detect the self-similar structure of the attractor, thus showing that the latter is a fractal set.

*References*: Falconer (1985); Kapitaniak (1991); Mandelbrot (1982).

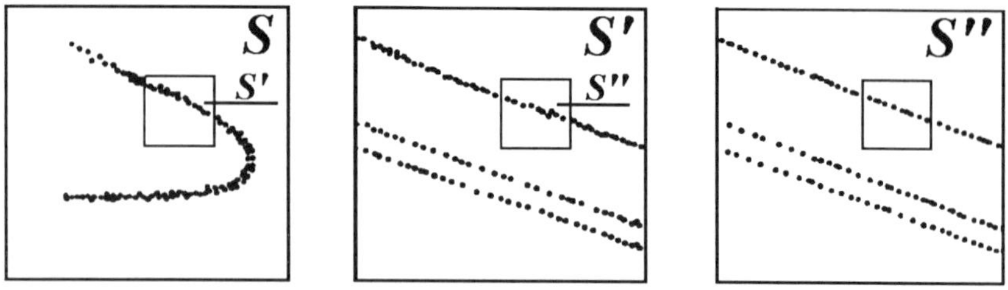


**Figure 55** Successive enlargements of the Poincaré map of a chaotic attractor

**frequency locking** (*see* **mode locking**)

**frequency response diagram**
This is a plot of the maximum amplitude of displacement versus the variation in frequency; it is also called a response curve.

**Frobenius–Perron operator (Perron–Frobenius operator)**
Let $(X, \mathcal{B}, \mu)$ be a measurable space. A map $S : X \to X$ is measurable if $S^{-1}(B) \in \mathcal{B}$ for all $B \in \mathcal{B}$.

A measurable map $S : X \to X$ on a measure space $(X, \mathcal{B}, \mu)$ is non-singular if $\mu(S^{-1}(B)) = 0$ for all $B \in \mathcal{B}$ such that $\mu(B) = 0$.

Let $(X, \mathcal{B}, \mu)$ be a measurable space. If $S : X \to X$ is a non-singular transformation, then the unique operator $P : L^1 \to L^1$, defined by

$$\int_B Pf(x)\mu(\mathrm{d}x) = \int_{S^{-1}(B)} f(x)\mu(\mathrm{d}x)$$

for all $B \in \mathcal{B}$, is called the Frobenius–Perron operator corresponding to $S$.

The Frobenius–Perron operator has the following properties:

(i) $P(\lambda_1 f_1 + \lambda_2 f_2) = \lambda_1 P f_1 + \lambda_2 P f_2$ for all $f_1, f_2 \in L^1$, and $\lambda_1, \lambda_2 \in \mathcal{R}$, so $P$ is a linear operator;

(ii) $Pf \geq 0$ if $f \geq 0$;

(iii) $\int_X Pf(x)\mu(\mathrm{d}x) = \int_X f(x)\mu(\mathrm{d}x)$;

(iv) if $S_n = S \circ \ldots \circ S$ and $P_n$ is the Frobenius–Perron operator corresponding to $S_n$, then $P_n = P^n$, where $P$ is the Frobenius–Perron operator corresponding to $S$.

*Example 1*: If $X = [a, b]$ is an interval on the real line $\mathcal{R}$ and $B = [a, x]$, then the Frobenius–Perron operator can be given in the explicit form

$$\int_a^x Pf(x)\mathrm{d}s = \int_{S^{-1}([a,x])} f(s)\mathrm{d}s$$

and by differentiating

$$Pf(x) = \frac{\mathrm{d}}{\mathrm{d}x} \int_{S^{-1}([a,x])} f(s)\mathrm{d}s.$$

*Example 2*: Let $S(x) = \exp x$. Thus, we have

$$Pf(x) = \frac{1}{x} f(\ln x).$$

*Reference*: Lasota and Mackey (1985).

# G

**Galerkin method**
This method is a procedure for the reduction of a set partial differential equation to a set of ordinary differential equations, used for continuous (i.e. in space and time) systems. Assuming a spatial variation and using the orthogonality of linear mode shapes, i.e. the solution is assumed in the form

$$u(x,t) = \sum_{n=1}^{N} u_n(t)\phi_n(x)$$

where $N \to \infty$, and where $\phi_n(x)$ are linear natural modes of the system, this method produces an approximate model based on coupled nonlinear ordinary differential equations.

*Reference*: Nayfeh and Mook (1979).

**Gauss map**
The map $x_{n+1} = f(x_n)$, given by

$$f(x) = \begin{cases} 0 & \text{if} \quad x = 0 \\ 1/x & \text{if} \quad x \neq 0 \end{cases}$$

is called the Gauss map.

*Remark*: This map preserves the Gauss measure

$$\mu(A) = \frac{1}{\ln 2} \int_A \frac{1}{1+x} \mathrm{d}x$$

where $A \subset [0,1)$ and integration in the Lebesgue sense is over the whole interval $[0,1)$.

**generic property**
The term generic is often used in nonlinear dynamics to mean typical in the sense that a generic property implies that small, continuous and differentiable

perturbations do not alter the system configuration. A strict mathematical definition of generic property is not easy to prescribe (see Jackson (1990) for more details).

*Reference*: Jackson (1990).

**geodesic flow**
Let $M$ be a compact connected smooth Riemannian manifold and define the sphere bundle $\Sigma$ as the set of all pairs $(m, \xi)$, where $m$ is an arbitrary point of $M$ and $\xi$ is a unit tangent vector starting at $m$ $(\Sigma = \{(m, \xi) : m \in M,\ \xi \in T_m, ||\xi|| = 1\})$. It can be proved that $\Sigma$, with an appropriately defined metric, is also a Riemannian manifold with a measure $\mu_\Sigma$. In a physical interpretation, $M$ is the configuration space of a system that moves with constant speed and $\Sigma$ is its phase space. Let $\gamma : \mathcal{R} \to M$ be a $C^1$ curve. The latter is called a geodesic if for every point $m_0 = \gamma(t_0)$ there is an $\epsilon > 0$ such that for every $m_1 = \gamma(t_1)$, with $|t_1 - t_0| \leq \epsilon$, the length of the arc $\gamma$ between the points $m_0$ and $m_1$ is equal to the distance between $m_0$ and $m_1$.

It can be proved that, for every $(m, \xi) \in \Sigma$, there exists exactly one geodesic satisfying

$$\gamma(0) = m, \quad \gamma'(0) = \xi, \quad ||\gamma'(t)|| = 1 \tag{1}$$

for all $t \in \mathcal{R}$.

Define a dynamical system $\{S_t\}$ , with $t \in \mathcal{R}$ on $\Sigma$, by

$$S_t(m, \xi) = (\gamma(t), \gamma'(t)) \tag{2}$$

where the geodesic $\gamma$ satisfies (1). This system is called a geodesic flow.

*Reference*: Lasota and Mackey (1985).

**giant squid axon** (*see* **axon**)

**Ginzburg–Landau equation**
The partial differential equation

$$\frac{\partial w}{\partial t} = \alpha w + \beta \frac{\partial^2 w}{\partial x^2} + \gamma w |w|^2$$

where $w$ is a complex-valued function of space and time, and the control parameters $\alpha, \beta$ and $\gamma$ are complex, is called the one-dimensional Ginzburg–Landau equation.

*Remark*: This equation is studied in connection with the spatio-temporal behaviour of fluids, superconductors, etc.

*Reference*: Drazin and Johnson (1989).

**global**
This term is applied to properties which cannot be analyzed in arbitrarily small neighbourhoods of a single point.

**global bifurcation**
Assume that the topological character of the phase portrait of the dynamical system changes when a control parameter is varied without changing the type of any fixed points of the system. Such a bifurcation is called a global bifurcation (*see also* **bifurcation**).

*Remark*: In the case of a global bifurcation the change in the phase portrait is not noticeable in the neighbourhood of any fixed point and can only be discerned on a global scale.

*Reference*: Guckenheimer and Holmes (1983).

**golden mean**
The number $\omega = 1/2(\sqrt{5} - 1)$ is called the golden mean.

*Remark*: The golden mean is considered as the most irrational number in the sense that it is the irrational number least easily approximated by rationals.

**gradient system**
Let $V : \mathcal{R}^n \to \mathcal{R}$ be any function. The $n$-dimensional system

$$\frac{\mathrm{d}x}{\mathrm{d}t} = -\nabla V(x)$$

is called a gradient system.

*Remark*: Gradient systems for which all fixed points are hyperbolic and all intersections of stable and unstable manifolds are transversal, are structurally stable.

*Reference*: Hirsch and Smale (1974).

**Grassberger–Procaccia dimension** (*see* **dimension of sets**)

**Grossman–Hartman (Hartman–Grossman) theorem for flows**
Let $A : \mathcal{R}^n \to \mathcal{R}^n$ be a linear transformation and let all eigenvalues $\lambda$ be such

that $|\lambda| \neq 1$. Every local diffeomorphism $B$, where

$$B : (\mathcal{R}^n, O) \to (\mathcal{R}^n, O)$$

with the linear part $A$ at the fixed point $O$, is topologically conjugate to $A$ in a sufficiently small neighbourhood of $O$.

*Remark*: In some texts this name is given as Grobman.

*References*: Arrowsmith and Place (1990); Irwin (1980); Steeb (1991).

# H

## Hamiltonian function

The Hamiltonian function $H$ is the function defined in terms of the Lagrangian function $L$, as follows:

$$H(p,q,t) = \sum_{i=1}^{n} p_i \frac{\mathrm{d}q_i}{\mathrm{d}t} - L(q, \frac{\mathrm{d}q}{\mathrm{d}t}, t) \tag{1}$$

where the coordinates $q = (q_1, ..., q_n)^T$ represent the generalised coordinates of an $n$-degree-of-freedom system and $\mathrm{d}q/\mathrm{d}t = (\mathrm{d}q_1/\mathrm{d}t, ..., \mathrm{d}q_n/\mathrm{d}t)^T$ are the generalised velocities. The canonical momenta $p_i$ are obtained from the Lagrangian according to

$$p_i = \frac{\partial L}{\partial(\mathrm{d}q_i/\mathrm{d}t)}. \tag{2}$$

The Lagrangian is obtained as the difference between the kinetic $T(\mathrm{d}q/\mathrm{d}t)$ and potential $V(q,t)$ energies in the usual way

$$L(q, \frac{\mathrm{d}q}{\mathrm{d}t}, t) = T\left(\frac{\mathrm{d}q}{\mathrm{d}t}\right) - V(q,t).$$

We use (2) to find $\mathrm{d}q_i/\mathrm{d}t = \mathrm{d}q_i(p,q)/\mathrm{d}t$ and thereby eliminate the dependence in (11) on the velocities in favour of a dependence on the canonical momenta and generalised coordinates.

## Hamiltonian system

A Hamiltonian system is a dynamical system given by

$$\frac{\mathrm{d}q}{\mathrm{d}t} = \frac{\partial H}{\partial p}$$

and

$$\frac{\mathrm{d}p}{\mathrm{d}t} = -\frac{\partial H}{\partial q}$$

where $p, q \in \mathcal{R}^{2n}$ and where $H$ is a Hamiltonian function.

If the Hamiltonian function does not depend on $t$, i.e. $H(p,q,t) = H(p,q)$, then the Hamiltonian system is said to be conservative.

*References*: Arnold (1988); Rasband (1990).

**harmonic balance method**
This is a method of approximation to periodic solutions of differential equations. If we assume that solutions are periodic with period $T = 2\pi/\omega$, then in line with a weighted residual approach a trial solution is considered of a form:

$$\bar{x} = a_0 + \sum_{i=1}^{N} a_{2i-1}\cos(i\omega t) + \sum_{i=1}^{N} a_{2i}\sin(i\omega t) \tag{1}$$

i.e. a Fourier type basis.

If, for example we consider solutions of the form (1) to the differential equation

$$\ddot{x} + f(x,\dot{x}) = \alpha\cos(\omega t + \phi) + \beta, \tag{2}$$

then substitution of (1) into (2) yields terms involving products of powers of $\sin(i\omega t)$ and $\cos(i\omega t)$. These latter terms can be expanded using trigonometric identities to leave all terms as sums of cosines and sines. We can now 'balance' the harmonic terms by equating to zero all the coefficients of cosines and sines (ignoring higher order terms), to yield expressions for the various coefficients in the approximation (1).

*References*: Hayashi (1964); Kapitaniak (1990).

**harmonic oscillator**
The dynamical system described by the linear equation

$$\frac{\mathrm{d}^2u}{\mathrm{d}t^2} + \Omega^2 u = 0$$

is called the harmonic oscillator; $\Omega$ is the frequency of the oscillations or natural frequency.

**Hartman–Grossman theorem** (*see* **Grossman–Hartman theorem**)

**Hausdorff dimension** (*see* **dimension of sets**)

**Hausdorff measure** (*see* **dimension of sets**)

**heat equation**
This is a partial differential equation which is used in the study of heat conduction and is given by

$$\frac{\partial u}{\partial t} = \alpha^2 \frac{\partial^2 u}{\partial x^2}$$

where typically $\alpha^2$ is the thermal diffusivity.

**Heighway's dragon**
This refers to a set in the plane. It is defined as a limit $P$ of the sequence of piecewise linear curves $P_n$, where $P_0$ is a line segment of length 1. $P_1$ results from $P_0$ by replacing the line segment by a piecewise linear curve with two segments of length $1/\sqrt{2}$, joined at a right angle. The two ends are the same as before. For $P_2$, each line segment in $P_1$ is replaced by a piecewise linear curve with two segments each having length $1/\sqrt{2}$ times the length of the segment that is replaced, etc. The first steps in the construction of Heighway's dragon are shown in Figure 56.

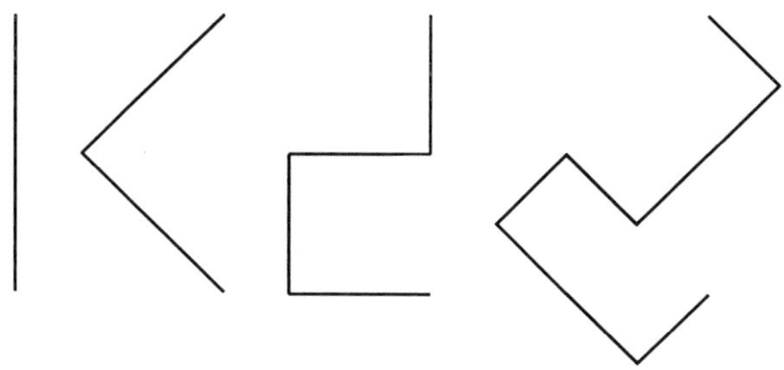

**Figure 56** First steps in the construction of Heighway's dragon

**Hénon–Heiles system**
The Hamiltonian system generated by

$$H(p,q) = \frac{1}{2}(p_1^2 + p_2^2) + \frac{1}{2}(q_1^2 + q_2^2) + q_1 q_2^2 - \frac{1}{3}q_1^3$$

is called a Hénon–Heiles system.

*Reference*: Hénon and Heiles (1964).

**Hénon map**
The two-dimensional map

$$x_{n+1} = y_n + 1 - a x_n^2$$

and

$$y_{n+1} = bx_n$$

where $a$ and $b$ are positive constants is called the Hénon map.

*Example*: For $a = 1.44$ and $b = 0.3$, iterations of this map appear to be chaotic, as shown in Figure 57.

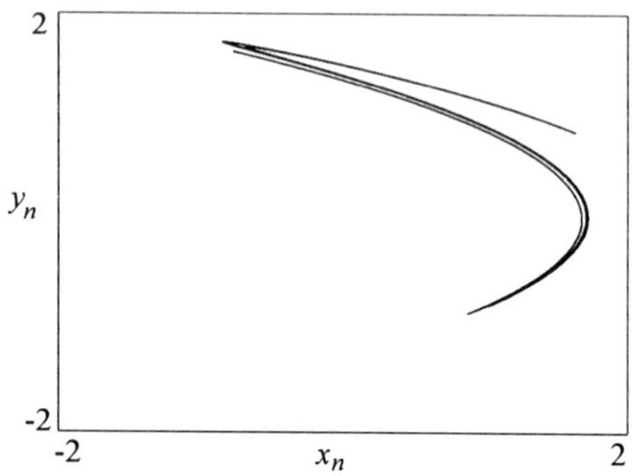


**Figure 57** Chaotic attractor of the Hénon map

*Reference*: Hénon (1976).

**heteroclinic point** (*see* **heteroclinic trajectory**)

**heteroclinic trajectory**

This is a trajectory between distinct points $u_1^*$ and $u_2^*$ which tends towards $u_1^*$ in reversed time and towards $u_2^*$ in forward time. Any point which maps $u_1^*$ in forwards and $u_2^*$ in backwards time is a heteroclinic point.

*Example*: Consider a mathematical pendulum

$$\frac{\mathrm{d}^2u}{\mathrm{d}t^2} + \sin u = 0$$

for $u \in [-\pi, \pi]$. It can be shown that the fixed point $(0, 0)$ is a centre and the fixed points $(-\pi, 0)$ and $(\pi, 0)$ are saddles. The phase-space portrait of the pendulum is shown in Figure 58.

Consider the orbit $\Gamma_1$, which starts at the unstable manifold $U_1$ of the saddle at $-\pi$ and finishes at the saddle at $\pi$ approaching it on the stable manifold $S_2$ of $\pi$. The orbit $\Gamma_2$, which starts at the unstable manifold of the saddle at $\pi$ of $U_2$ and connects to the other saddle via $S_1$, has a similar property.

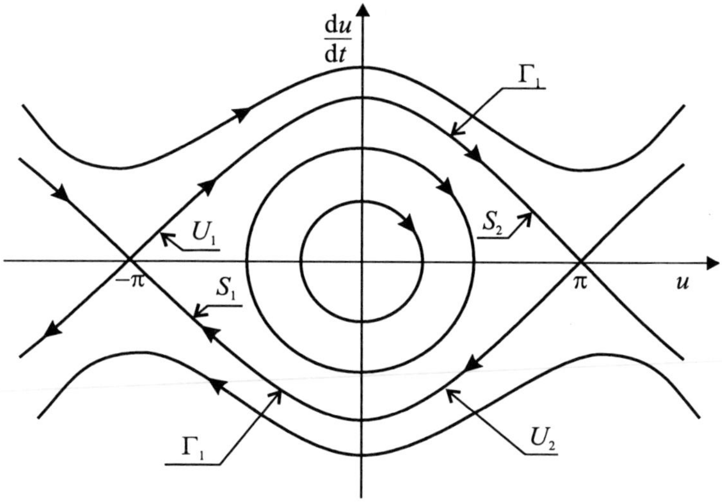


**Figure 58** Phase-space portrait of a pendulum

The orbit $\gamma \subset \Gamma_1 \cup \Gamma_2$ is called a *heteroclinic orbit.*

**Hilbert's 16th problem**

This problem concerns the number, and stability properties, of limit cycles for systems described by

$$\frac{\mathrm{d}x}{\mathrm{d}t} = F(x, y)$$

and

$$\frac{\mathrm{d}y}{\mathrm{d}t} = G(x, y).$$

Even when $F$ and $G$ are simple polynomials, the solution to Hilbert's 16th problem, remains an open question.

*Reference*: Drazin (1992).

**Hill's equation**

An ordinary differential equation of the form

$$\frac{\mathrm{d}^2x}{\mathrm{d}t^2} + P(t)x = 0$$

where $P(x)$ is a periodic function, is called Hill's equation.

Hill's equation was derived as an approximation to the 3-body problem.

**hill-top saddle**
The term hill-top saddle refers to the global unstable saddle-type solution around a global maximum in the governing potential energy function in problems relating to escape from a potential well.

*Example*: Consider the dynamics of the unforced Thompson–Helmholz escape equation

$$\frac{d^2x}{dt^2} + b\frac{dx}{dt} + x - x^2 = 0.$$

This system has a potential function $V(x)$, as shown in Figure 59(a), with a hill-top saddle global maximum ($x = 1$) which corresponds to an unstable fixed point.

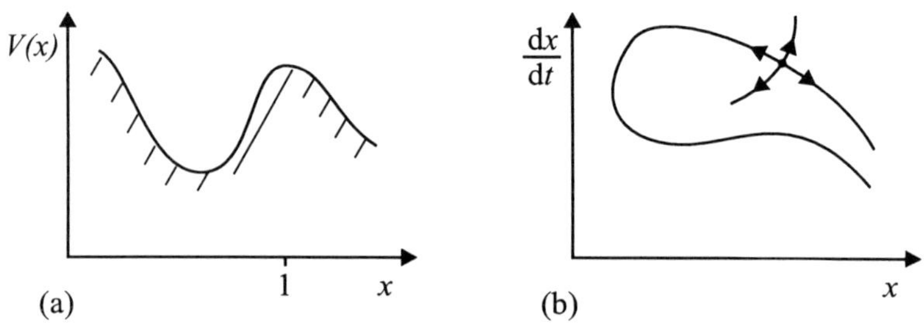


**Figure 59** (a) Potential function, and (b) Poincáre map of Thompson–Helmholz escape equation

When small forcing is applied, then this point may be associated with a small unstable periodic orbit. When the system is stroboscopically sampled, then this solution produces a single point which corresponds to a saddle point in the Poincaré phase plane, as shown in Figure 59(b).

**Hodgkin–Huxley equation**
The fundamental equations of electrophysiology, namely the prototype for an exact description of the dynamics of voltage and of specific ion conductivities in cell membranes.

*Reference*: Winfree (1988).

**Holling–Tanner model**
This is a model for population dynamics of the prey-predator type which involve oscillations which are structurally stable. It is described by the

following equations:

$$\frac{\mathrm{d}x_1}{\mathrm{d}t} = r\left(1 - \frac{x_1}{k}\right)x_1 - \frac{wx_1x_2}{(D + x_1)}$$

and

$$\frac{\mathrm{d}x_2}{\mathrm{d}t} = s\left(1 - \frac{\tau x_2}{x_1}\right)x_2$$

with $r, s, k, D, and \tau > 0$.

*Reference*: Arrowsmith and Place (1990).

**homeomorphism**
This is a one-to-one and onto map $f : X \to Y$ for which both $f$ and inverse $f^{-1}$ are continuous.

*Reference*: Jackson (1990).

**homoclinic point** (*see* **homoclinic trajectory**)

**homoclinic tangle**
Let $M$ be a compact two-dimensional manifold and $f$ be $C^1$ diffeomorphism on $M$ with the hyperbolic saddle point $x_0$. If the stable $W^s$ and unstable $W^u$ manifolds of the hyperbolic saddle point $x_0$ intersect at some point $x_1$, then they must intersect infinitely many times. Figure 60 illustrates the effect that this constraint has on the two manifolds if we attempt to return them directly to $x_0$ itself. As the unstable manifold approaches the saddle point the loops between adjacent homoclinic points are stretched parallel to $W^s$ and squeezed parallel to $W^u$. The manifold therefore undergoes oscillations of increasing amplitude and decreasing period. The fate of the stable manifold is similar under reverse iterations, thus resulting in the homoclinic tangle shown in Figure 60. The parallelogram $R$ has images $R_1 = f^N(R)$ and $R_0 = f^{-N}(R)$, intersecting in a horseshoe configuration.

**homoclinic trajectory**
This is a trajectory of a point $u^*$ which tends to $u^*$ in both forward and reverse time. Each point which tends to $u^*$ is called a homoclinic point.

*Example*: Consider the equation

$$\frac{\mathrm{d}^2u}{\mathrm{d}t^2} + bu + cu^3 = 0$$

where $b < 0$ and $c > 0$ are constant. The homoclinic orbit to the fixed point

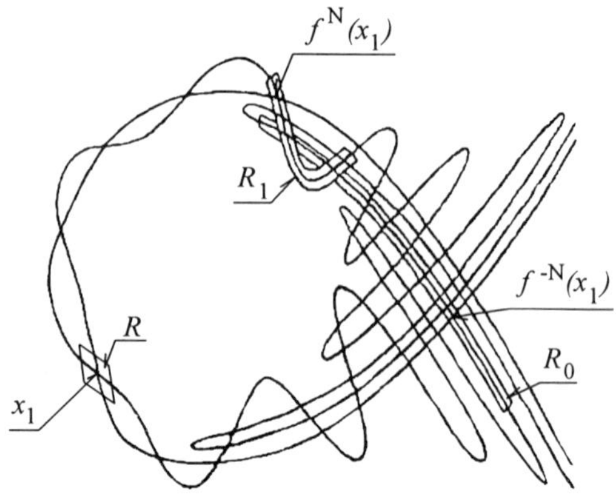


**Figure 60** The homoclinic tangle occuring at a hyperbolic saddle point

(0, 0) is given by

$$u(t) = \sqrt{\frac{-2b}{c}} \sec h(\pm(-1)^{1/2}(t - t_0))$$

and

$$\frac{\mathrm{d}u(t)}{\mathrm{d}t} = -b\sqrt{\frac{2}{c}} \sec h(\pm(-1)^{1/2}(t - t_0) \tanh(\pm(-b)^{1/2}(t - t_0)).$$

*References*: Kapitaniak (1991); Wiggins (1990).

**homogeneous function**
The function $P(x, \lambda)$, where $\lambda \in \mathcal{R}$ is a homogeneous function of degree $n$ $(n = 1, 2, ...)$ if $P(x, \lambda) = \lambda^n P(x)$.

**homotopy**
This theory is the branch of topology which deals with the continuous deformations of sets.

Let $X$ and $Y$ be two topological spaces, and $f$ and $g$ two continuous maps $f : X \to Y$, and $g : X \to Y$. Then $f$ is said to be homotopic to $g$ if there exists a map $h : X \times [g] \to Y$, continuous in $(x, t)$ such that $h(x, 0) = f$ and $h(x, 1) = g$.

Two spaces $X$ and $Y$ are called homotopically equivalent if there exist maps $f : X \to Y$ and $g : Y \to X$ such that the composite maps $fg : Y \to Y$ and

$gf : X \to X$ are homotopic to their respective identity maps $I : Y \to Y$ and $I : X \to X$.

**Hopf bifurcation**
The Poincaré–Andronov–Hopf bifurcation, which is frequently simply called the Hopf bifurcation, involves the change in stability of a focus in a dynamical system

$$\frac{dx}{dt} = f(x, c) \tag{1}$$

where $x \in \mathcal{R}^2$, and $c \in \mathcal{R}$, as a control parameter $c$ is varied, together with the birth of a periodic orbit, as illustrated in Figure 61. The Hopf bifurcation can occur in any (larger than one) dimensional system and can be studied via a two-dimensional centre manifold.

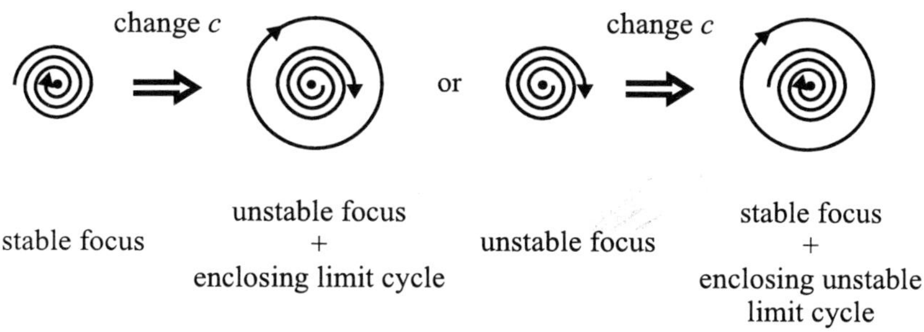


**Figure 61** Generic Hopf bifurcations

The bifurcations can be viewed more completely in the response-control space, where the bifurcation point is taken to be $c = 0$. In this space, one can see that there is a surface which contains only periodic solutions of the nonlinear equation (1). This surface is called the centre manifold (*see* **centre manifold theorem**) in the response-control space.

Periodic solutions are either stable or unstable limit cycles. If the surface consists of stable limit cycles, as in Figure 62(a), then the bifurcation is said to be supercritical (or soft excited, since if $c$ is only slightly positive, the motion away from the fixed point is very small since there is a nearby stable limit cycle). If the centre manifold consists of unstable limit cycles ) (see Figure 62(b)), the bifurcation is referred to as subcritical (or hard excited; as a small perturbation can take the state 'outside' the unstable limit cycle; the trajectory will then continue to some distant part of the phase space).

A degenerate Hopf bifurcation is a bifurcation in which as a parameter moves

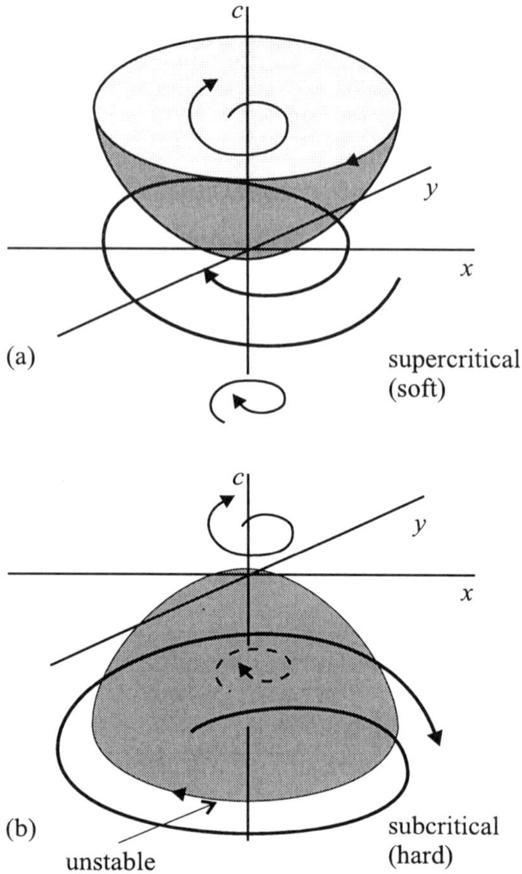


**Figure 62** (a) Supercritical and (b) subcritical Hopf bifurcations

through the bifurcation point the equilibrium position changes from a stable spiral to an unstable spiral but for which no unique periodic orbit exists.

*Reference*: Jackson (1990).

**Hopf theorem**

Let $G$ be an open connected domain in $\mathcal{R}^n$, with $\mu > 0$, and let $F$ be a real analytic function defined on $G \times [-\mu, \mu]$. Consider the differential system

$$\frac{\mathrm{d}x}{\mathrm{d}t} = F(x, c), \tag{1}$$

where $x \in G, |c| < \mu$. Suppose that there is a real analytic, vector function $g$ defined on $[-\mu, \mu]$ such that $F(g(c), c) = 0$. Expand $F(x, c)$ about $g(c)$ to form

$$F(x,c) = L_c\bar{x} + F^*(\bar{x},c), \tag{2}$$

in which $\bar{x} = x - g(c)$, and where $L_c$ is an $n \times n$ real matrix which depends only on $c$, and $F^*(\bar{x},c)$ is the nonlinear part of $F$. Suppose that there exist exactly two complex conjugate eigenvalues, $\alpha(c)$ and $\bar{\alpha}(c)$, of $L_c$ with the properties

$$\mathrm{Re}\,((\alpha(0)) = 0$$

and

$$\mathrm{Re}\left(\frac{\mathrm{d}\alpha(0)}{\mathrm{d}c}\right) \neq 0$$

Then there exists a periodic solution $P(t,\epsilon)$ with period $T(\epsilon)$ of (1) with $c = c(\epsilon)$, such that $c(0) = 0$, $P(t,0) = g(0)$ and $P(t,\epsilon) \neq g(c(\epsilon))$ for all sufficiently small $\epsilon \neq 0$. Moreover, $c(\epsilon)$, $P(t,\epsilon)$, and $T(\epsilon)$ are analytic at $\epsilon = 0$, and $T(0) = 2\pi/|\mathrm{Im}\ \alpha(0)|$. These 'small' periodic solutions exist for exactly one of three cases, namely either only for $c > 0, c < 0$, or $c = 0$.

*Reference*: Chow and Hale (1982).

**Hopfield model**
This is a model of a neural network, which can be formed by an electric resistance–capacitance (RC) circuit connecting amplifiers. The input voltage $u$ and output voltage $V$ are characterised by the function $V = g(u)$, as shown in Figure 63.

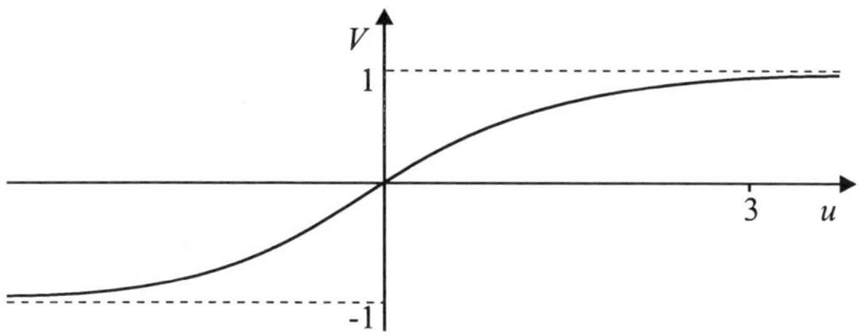


**Figure 63** Relationship between input $u$ and output $V$ voltages in the Hopfield model

*Reference*: Drazin (1992).

**Hopf–Landau theory (of hydrodynamic turbulence)** (*see* **Landau–Hopf theory**)

**horseshoe map**

This map is specified geometrically in Figure 64. The map takes the square $S$, uniformly stretches it vertically by a factor $\mu > 2$, and then uniformly compresses it horizontally by a factor $\lambda \in (0, 1/2)$. The long strip is next bent into a horseshoe shape with all the bending deformations taking place in the uncrossed–hatched regions of the Figure. The horseshoe (the transformed set $f(S)$) is placed on the top of the original square, so that the map is now confined to a subset of the original unit square. If the entire sequence of operations is repeated, then four stripes appear from the original two and so on. The repetition of the process $n$ times leads to $2^n$ stripes, and a cut across the stripes would, in the limit of large $n$, lead to a fractal, or more precisely, a Cantor set. Note that in each iteration a certain fraction of the original area of the square $S$ is mapped to the region outside the square, and in the limit $n \to \infty$ almost every initial condition with respect to Lebesque measure leaves the square, so that the Cantor-like limit set is dynamically unstable.

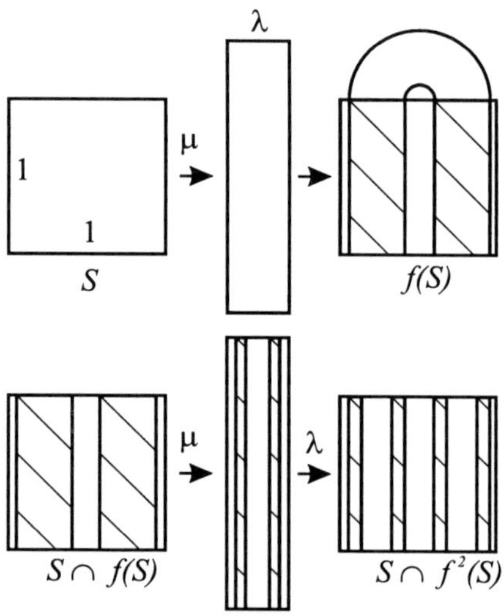


**Figure 64** Construction of the horseshoe map (two iterations)

A configuration similar to the horseshoe map occurs in the phase space of dynamical systems where there are regions with strong contraction and expansion. For example, in the neighbourhood of saddle points, trajectories approach fixed points most rapidly along the stable manifold, and depart most rapidly along the unstable manifold.

Tangent vectors along the stable manifold are contracting (negative Lyapunov exponents), and tangent vectors along the unstable manifold are expanding (positive Lyapunov exponent). Any region of phase space where these two types of behaviour are in close proximity may exhibit stretching and folding.

*Example*: Consider a damped unforced pendulum modelled by the equation

$$\frac{d^2x}{dt^2} + a\frac{dx}{dt} + \sin x = 0.$$

After a fixed-point analysis, one arrives at the phase plane shown in Figure 65. The stable and unstable manifolds of the saddle-type fixed point at $x = \pm\pi$, where $dx/dt = 0$, are the trajectories that approach and depart most quickly from the unstable fixed points.

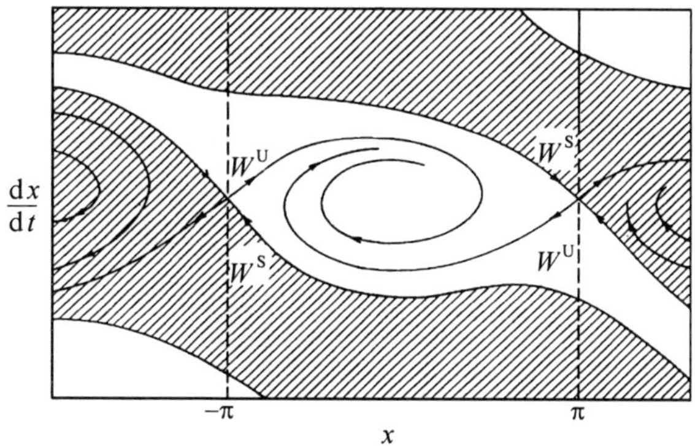


**Figure 65** Phase plane of the damped unforced pendulum

Now consider the lightly driven pendulum

$$\frac{d^2x}{dt^2} + a\frac{dx}{dt} + \sin x = f\cos t$$

where $f << 1$. In this case, the plot of Figure 65 can be approximately considered as a Poincaré map of the three-dimensional phase space, except that the lines should be regarded as a sequence of dots (mapping points) corresponding to successive passages of the trajectories through the cross-section.

When the forcing is increased, the stable and unstable manifolds can cross, for example at the point $I_1$. We can observe that each crossing is mapped into another one which is closer to the saddle point, thus leading to an infinite number of intersections, $I_2, I_3, ...$, etc. (Figure 66).

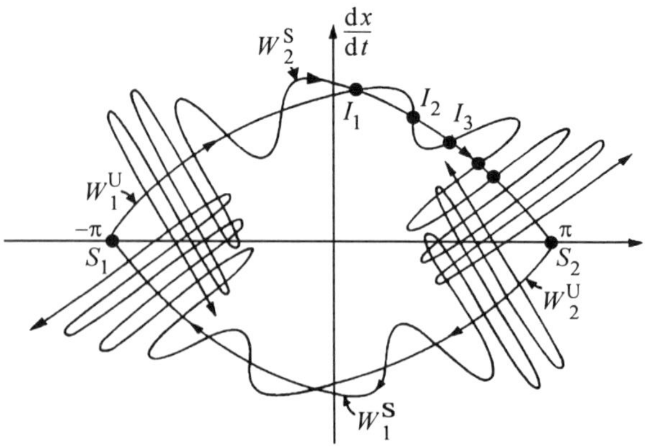


**Figure 66** Intersection of stable $W^s$ and unstable $W^u$ manifolds on the Poincaré map of the forced pendulum

A small rectangular section of the Poincaré map near $I_1$ undergoes stretching and folding much like that of the horseshoe map, due to the strong bending of the manifolds near the saddle point.

*Remark*: Generally, it is shown that the intersection of stable and unstable manifolds (for example, between $I_1$ and $I_2$) is topologically equivalent to (i.e. can be smoothly transformed into) the interated Smale horseshoe. This horseshoe configuration of the Poincaré map shows that two points which are initially close together will move apart after a few iterations. Therefore, a horseshoe map can be considered as a prototype of chaotic dynamics.

*References*: Newhouse (1980); Wiggins (1988).

**hyperbolic fixed point**
Let $u^*$ be a fixed point of

$$\frac{\mathrm{d}u}{\mathrm{d}t} = f(u)$$

where $u \in \mathcal{R}^n$ and $A$ is the matrix $A_{i,j} = (\partial f_i)/\partial x_i|_{u^*}$. Then $u^*$ is a hyperbolic point if $A$ has no purely imaginary eigenvalues.

**hyperbolic set**
Let $M$ be a compact smooth $m$-dimensional manifold with $T_xM$ the tangent space to $M$ at the point $x$, and let $f : M \to M$ be a diffeomorphism. Let $\Lambda$ be a closed subset of $M$, invariant under the diffeomorphism $f$. Assume that we have linear subspaces $E_x^- and E_x^+$ of $T_xM$ for each $x \in \Lambda$, depending

continuously on $x$ such that

$$T_x M = E_x^- + E_x^+$$

and

$$\dim E_x^- + \dim E_x^+ = m$$

and that

$$T_x f E_x^- = E_{fx}^-$$

furthermore

$$T_x f E_x^+ = E_{fx}^+.$$

$E_x^-$ and $E_x^+$ form a continuous invariant splitting of $TM$ over $\Lambda$. The subset $\Lambda$ is called a hyperbolic set if there exist constants $C > 0$ and $\Theta > 0$ such that , for all $n \geq 0$

$$||T_x f^n u||_{f^n x} \leq C\Theta^{-n}||u||_x$$

where $||.||$ is a norm in $T_x M$ if $u \in E_x^+$, and

$$||T_x f^{-n} v||_{f^n x} \leq C\Theta^{-n}||v||_x$$

if $v \in E_x^-$.

**hyperbolic system** (*see* **Axiom-A diffeomorphism, Axiom-A flow**)

**hyperchaos**
Let the $n$-dimensional dynamical system have an orbit with a spectrum of Lyapunov exponents, $\lambda_1 \geq \lambda_2 \geq \lambda_3 \geq .... \geq \lambda_n$. If

$$\sum_{i=1}^{n} \lambda_i < 0$$

and at least two Lyapunov exponents are positive, then the orbit evolves on a hyperchaotic attractor. This type of behaviour is referred to as hyperchaos.

*Reference*: Rössler (1983).

**hysteresis**
This is the term used to represent global behaviour whereby a function or system response is different for increasing or decreasing parameter values.

*Example*: Consider a vibrating mass whose motion is governed by a softening nonlinear spring, as shown in Figure 67(a).

A response-curve plot of the maximum amplitude of the displacement, $|x|$, shown in Figure 67(b), bends over to the left so that for increasing $\omega$ the

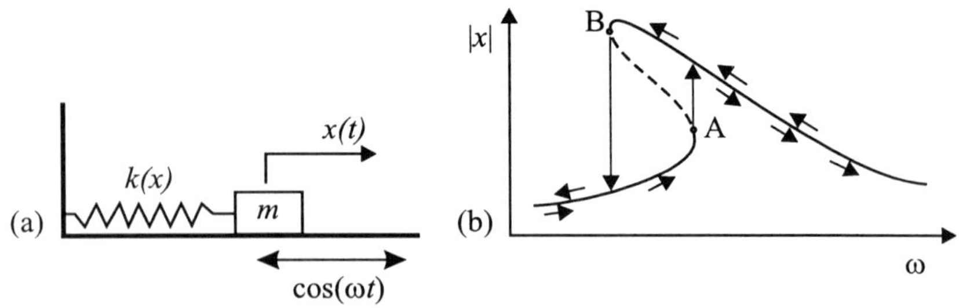


**Figure 67** (a) Model of a vibrating system, and (b) the corresponding response curve

system jumps at $A$ to the resonant motion, while for decreasing $\omega$ the system would evolve along the resonant branch to the point $B$ where a jump to the non-resonant motion would occur.

# I

**Ikeda map**
Let $z$ be a complex number, $z = x + iy$, where $x, y \in \mathcal{R}$. The map given by the relationship

$$z_{n+1} \to r + c_2 z\, exp\left[i\left(c_1 - \frac{c_3}{1+|z|^2}\right)\right]$$

where $r, c_{1,2,3}$ are real parameters, is called the Ikeda map.

*Example*: For $r = 0.85$, $c_1 = 0.4$, $c_2 = 0.9$ and $c_3 = 7.2789$, this map shows chaotic behaviour, as illustrated in Figure 68.

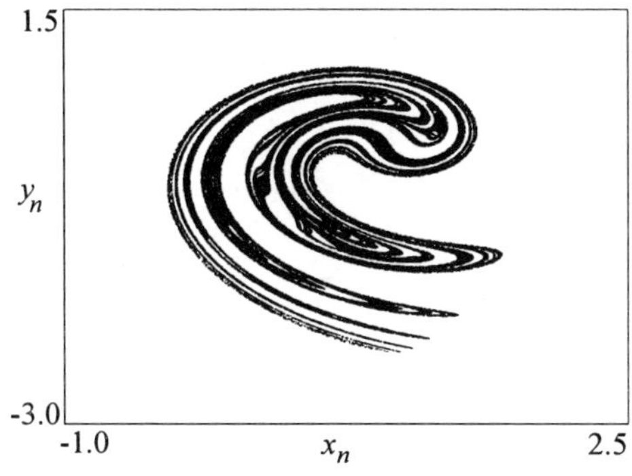


**Figure 68** Chaotic behaviour of the Ikeda map

**impact oscillator**
An the name implies, this describes an oscillator with impacts. Consider a mass moving along $x$ which is driven, but also undergoes repeated impacts caused by a motion-limiting constraint at $x = a$. The simplest form considers the motion to be linear between impacts and may be modelled by

$$\frac{\mathrm{d}^2 x}{\mathrm{d}t^2} + b\frac{\mathrm{d}x}{\mathrm{d}t} + x = F\cos\omega t$$

for $x < a$ and

$$x^+ = rx^-$$

for $x = a$, where $x^+$ and $x^-$ denote the velocity before and after impact, respectively, and $r$ is the coefficient of restitution. Such systems are also referred to as vibro-impact systems.

*References*: Foale and Bishop (1994); Kobrinskii and Kobrinskii (1973); Nordmark (1991); Shaw and Holmes (1982).

**imperfect bifurcation**
Bifurcations which occur due to the imperfection of physical materials are referred to as imperfect bifurcations.

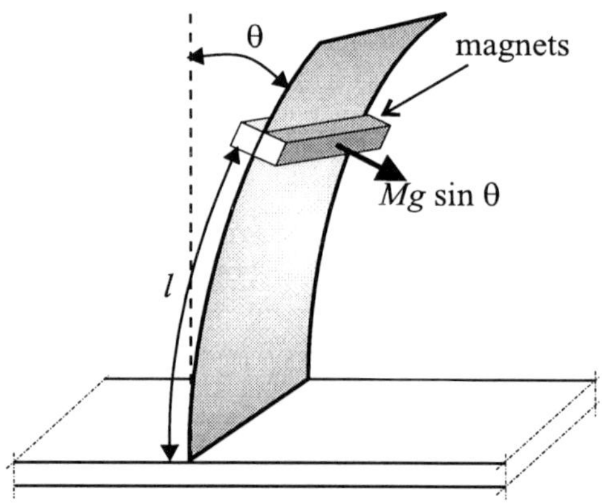


**Figure 69** A strip with two magnets

*Example*: Consider a light, stiff metal strip (shown in Figure 69) with its bottom edge attached to a table, and two magnets on opposite sides of the strip (to maintain symmetry), which can be moved up or down the strip at a distance $l$ from the bottom (used here as a control parameter). The strip is initially vertical with the weights at the bottom ($l = 0$). Assume that the torque generated by the strip about the bottom axis, when it is displaced by an angle $\theta$, is simply proportional to $\theta$, i.e. the *strip's torque* $= -\eta(l)\theta$, where $\eta$ is the torsional stiffness of the strip. Newton's equation, when the strip experiences a damping force $\mu \mathrm{d}\theta/\mathrm{d}t$, is

$$Ml^2 \frac{\mathrm{d}^2\theta}{\mathrm{d}t^2} = -\mu \frac{l\mathrm{d}\theta}{\mathrm{d}t} - \eta\theta + Mgl\sin\theta$$

where $g$ is the gravitational acceleration, and ignoring the inertial effect

$$|\mu d\theta/dt| \gg |Ml^2 d^2\theta/dt^2|)$$

which gives the approximate equation

$$\frac{\mathrm{d}\theta}{\mathrm{d}t} = -\frac{\eta}{\mu}\theta + \frac{MG}{\mu} l \sin\theta = F(\theta, l).$$

The equilibrium states are then given by

$$\theta = \frac{l}{D}\sin\theta, \tag{1}$$

where $l = \eta/mg$. One readily concludes that $l = D$ is the bifurcation point, as shown in Figure 70(a) where the ship buckles. The symmetry is broken if the strip is not orientated vertically (due to the physical imperfections) when $l = 0$. In this case (1) is replaced by

$$\theta = \frac{l}{D}\sin(\theta + \phi) \tag{2}$$

and this leads to the imperfect bifurcation shown in Figure 70(b).

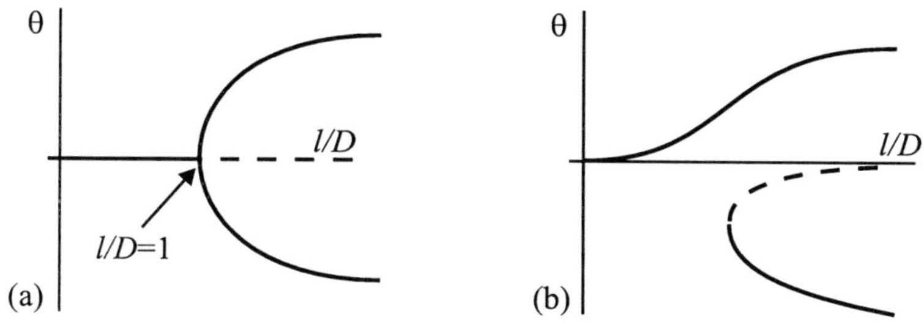


**Figure 70** (a) Ideal and (b) imperfect bifurcation diagrams

*References* Jackson (1990).

**implicit function theorem**
Let $F(x, y) = 0$ be a system of $n$ equations, and let $A \in \mathcal{R}^n$ and $B \in \mathcal{R}^m$ be regions such that, for $x \in A$ and $y \in B$, this system has a real solution, and the determinant of the Jacobian does not vanish, i.e.

$$\det\left|\frac{\partial F_i}{\partial x_j}\right| = \left|\frac{\partial(F_1, \ldots, F_n)}{\partial(x_1, \ldots, x_n)}\right| \neq 0.$$

Then there is a region $C \subset B$, such that for $y \in C$ there is a unique solution

$$x = g(y)$$

for $x \in A$, which is continuous in $y$, such that

$$F(g(y), y) = 0$$

for $y \in C$.

*Remark*: Note that the size of the region $C$ is not specified, only that it exists. Therefore the unique inversion may only be local.

*References*: Guckenheimer and Holmes (1983); Hirsch and Smale (1974); Jackson (1990).

**improper node** (*see* **fixed points**)

**indeterminate bifurcation**
This is the term used when a bifurcation occurs for which the basin boundaries are fractal, such that for very small changes in a control parameter the system may jump from a steady state to any of a number of co-existing solutions.

*Reference*: Thompson (1992).

**inflected node** (*see* **fixed point**)

**information dimension** (*see* **dimension of sets**)

**inset**
Another name for a stable manifold (*see* **invariant set**).

**integrable**
Let $H$ be a Hamiltonian function

$$H(p, q) = \sum_{i=1}^{N} \frac{p_i^2}{2m_i} + U(q)$$

with $N$ degrees of freedom. If the $N$ first integrals in involution exist with linearly independent gradients (*see* **action-angle variables**), then the Hamiltonian system is said to be integrable.

*Remark*: One of the first integrals is the Hamiltonian function itself. Two first integrals, $I_1$ and $I_2$, are in involution if $\{I_1, I_2\} = 0$, where $\{.\}$ is the Poisson bracket.

*Example*: Consider the Hamiltonian function

$$H(p, q) = \frac{1}{2}(p_1^2 + p_2^2) + e^{q_2 - q_1}.$$

Then a first integral, in addition to the Hamiltonian function, is

$$I(p) = p_1 + p_2.$$

*Reference*: Steeb (1991).

**integrals of motion** (*see* **constants of the motion**)

**interior**
The interior of a set $A$, $int\ A$, is the largest open set contained in $A$.

**intermingled basins** (*see* **riddled basins**)

**intermittency**
By intermittency we mean the occurrence of fluctuations that alternate 'randomly' between long periods of regular behaviour and relatively short irregular bursts, i.e. the motion is nearly periodic with occasional irregular bursts. Typical time historiy of such intermittent transitions are shown in Figure 71.

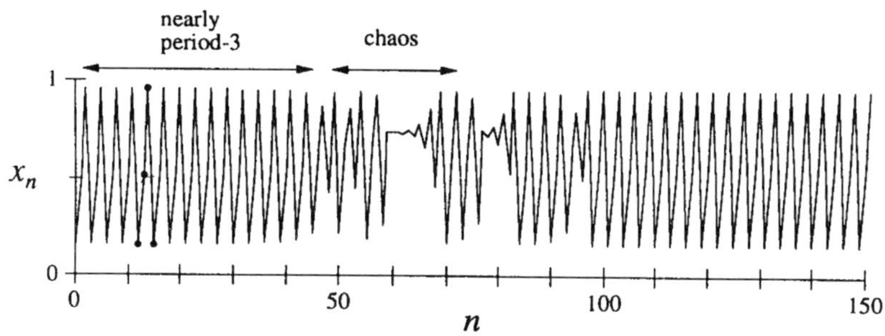


**Figure 71** Typical time series of intermittent behaviour

It has been found that the density of chaotic bursts increases with an external control parameter, which shows that intermittency presents a continuous route from regular to chaotic behaviour.

There are three types of intermittency corresponding to the three types of generic bifurcation of one-dimensional maps (*see* **bifurcation**). The first type is characterised by the loss of stability when the real eigenvalue of the fixed point crosses the unity circle at +1 and a tangent bifurcation occurs. The second type is characterised by the simultaneous crossing of the two complex eigenvalues of the unit circle, i.e. the system undergoes subcritical Hopf bifurcation. When the real eigenvalue crosses the unit circle at $-1$, i.e. the system undergoes subcritical period-doubling bifurcation, this corresponds to a third type of intermittency.

*Reference*: Pomeau and Manneville (1980).

**invariant measure**
A measure $\mu$ is said to be invariant for the map $f : \mathcal{R}^n \to \mathcal{R}^n$, if for any set $X \subset \mathcal{R}^n$

$$\mu(X) = \mu(f^{-1}X)$$

where $f^{-1}X$ are all of the points that map on to $X$.

**invariant set (manifold)**
Let $A \subset \mathcal{R}^n$ and $f : \mathcal{R}^n \to \mathcal{R}^n$. Then the set $A$ is an invariant set if $f(A) = A$.

*Example*: Fixed points are invariant sets.

An invariant set which is also a manifold is called an invariant manifold.

A stable manifold is also sometimes known as the inset.

*Reference*: Thompson and Stewart (1986).

**inverse map**
The inverse of the map $f : X \to Y$ at $y \in Y$, is the collection of all of the points $x$ in $X$ such that $y = f(x)$.

*Remark*: $f^{-1} : Y \to X$ is a map only if $f(X)$ is both one-to-one and onto.

**invertible map**
A map whose inverse exists and is unique, is said to be invertible (*see* **inverse map**).

**iterated function system**
This consists of a complete metric space $(X, d)$ together with a finite set of contraction maps

$$f_n : X \to X$$

with respective contractivity factors $s_n$, for $n = 1, 2, ..., N$. The notation for the iterated function system is

$$\{X; f_n, n = 1, 2, ..., N\}$$

and its contractivity factor is

$$s = \max\{s_n : n = 1, 2, ..., N\}.$$

*Reference*: Barnsley (1988).

**Ito equation**
Let $e$ and $f$ be two functions from $[0, T] \times \Omega$ to $\mathcal{R}$, where $\Omega$ is a set of elementary events, with such measurability and integrability properties that the ordinary and stochastic integrals appearing in the following formula (1) make sense.

By the stochastic differential Ito equation we mean an expression

$$\mathrm{d}X_t(\omega) = e(t, \omega)\mathrm{d}t + f(t, \omega)\mathrm{d}W_t(\omega) \tag{1}$$

where $W_t(\omega)$ is a Wiener stochastic process. Expression (1) is a symbolic way of writing

$$X_t(\omega) - X_s(\omega) = \int_s^t e(u, \omega)\mathrm{d}u + \int_s^t f(u, \omega)\mathrm{d}W_t(\omega)$$

with probability 1, for any $0 \leq s \leq t \leq T$.

*Reference*: Kloeden and Platen (1992).

# J

## Jacobi elliptic functions

These elliptic functions are the inverses of elliptic integrals (*see* **elliptic integrals**). There are five elliptic functions, defined as follows:

$$sn^{-1}(x,k) = \int_0^x \frac{dy}{[(1-y^2)(1-k^2y^2]^{1/2}} = F(x,k)$$

and

$$cn^{-1}(x,k) = \int_x^1 \frac{dy}{[(1-y^2)(k'+k^2y^2]^{1/2}} = F([1-x^2]^{1/2},k)$$

for $(0 \leq x \leq 1)$ and $k' = 1-k^2$;

$$dn^{-1}(x,k) = \int_x^1 \frac{dy}{[(1-y^2)(k'^2+k^2y^2]^{1/2}} = F([1-x^2]^{1/2}/k^2,k)$$

for $(k' \leq x \leq 1)$;

$$tn^{-1}(x,k) = \int_0^x \frac{dy}{[(1-y^2)(1+k'^2y^2]^{1/2}} = F([x^2/(1+x^2]^{1/2},k)$$

for $(0 \leq x \leq \infty)$;

$$am^{-1}(\phi,k) = \int_0^\phi \frac{d\theta}{[1-k^2\sin^2\theta]^{1/2}} = F(\sin\phi,k).$$

Jacobi elliptic functions, as well as elliptic integrals, frequently appear as the solutions of particular nonlinear differential equations.

*Reference*: Byrd and Friedman (1971).

## Jacobian matrix

Let $f : A \rightarrow \mathcal{R}^n$ be a function, where $A$ is an open subset of $\mathcal{R}^m$. Assume that the component functions possess first-order partial derivatives. The matrix

$$\left(\frac{\partial f_j}{\partial x_k}\right)$$

where $j = 1, ..., n$ and $k = 1, ..., m$, is the Jacobian matrix of $f$.

When $n = m$, the determinant of the Jacobian matrix is called the Wronskian.

**Japanese attractor**
A term used by David Ruelle to describe the chaotic attractor of Ueda's equation:

$$\frac{d^2x}{dt^2} + 0.1\frac{dx}{dt} + x^3 = 10\cos t.$$

(*see* **Ueda's equation**).

**Jordan form (Jordan normal form)**
The Jordan normal form of an $(n \times n)$-dimensional matrix $A$ is a matrix $B = PAP^{-1}$, where $P$ is an invertible $(n \times n)$-dimensional matrix with the non-zero entries of $B$ in the form of diagonal blocks:

$$\begin{pmatrix} \lambda_1 & 1 & \cdots & 0 \\ 0 & \lambda_{-2} & 1 & \cdots \\ \cdots & \cdots & \cdots & \cdots \\ 0 & \cdots & \cdots & \lambda_n \end{pmatrix}.$$

*Reference*: Hirsch and Smale (1974).

**Josephson junction array**
A Josephson junction is a superconducting device capable of generating high frequency voltage oscillations. Mechanical analogies result in an equation of the form of a driven pendulum:

$$\frac{d^2\phi}{dt} + b\frac{d\phi}{dt} + b\sin\phi = F$$

where $b$ is constant and $F$ is an applied torque.

*Reference*: Strogatz (1994).

**Julia set**
Let $C_\lambda(z)$ be a mapping in the complex plane. The Julia set of $C_\lambda$, denoted $J(C_\lambda)$, is the closure of the set of repelling periodic points.

*Remark*: A complex-valued point $z_0 \in C$ is periodic if $C_\lambda^n(z_0) = z_0$ for some $n$, where $C_\lambda^n$ denotes the $n$-fold composition of $C_\lambda$ with itself. This periodic point is repelling if $|(C_\lambda^n)'(z_0)| > 1$.

*Example*: The Julia set for the map $f(z) = z^2 + c$, where $c = -0.12256117 + 0.74486177i$, is shown in Figure 72.

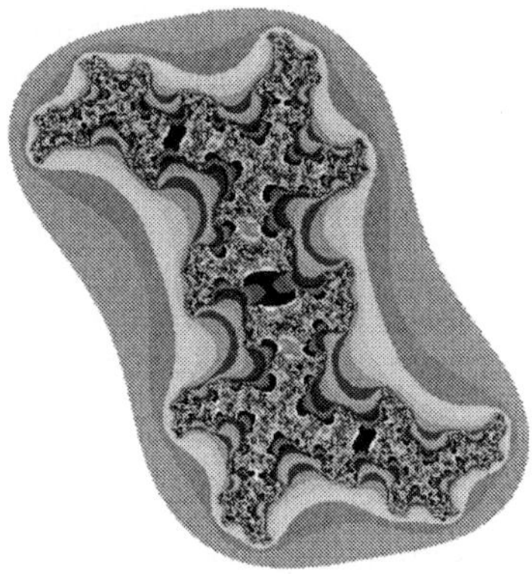

**Figure 72** Representation of a Julia set

*References*: Barnsley (1988); Steeb (1991).

**jump phenomenon**
Consider a periodically forced system

$$\frac{\mathrm{d}^2x}{\mathrm{d}t^2} + \omega^2 x + f\left(x, \frac{\mathrm{d}x}{\mathrm{d}t}\right) = A\cos\Omega t \tag{1}$$

where $A$, $\omega$ and $\Omega$ are constants. Assume that equation (1) has an approximate periodic solution, $C\cos(\Omega t + \phi)$. The plot $C$ versus $\Omega$ is called the resonance curve.

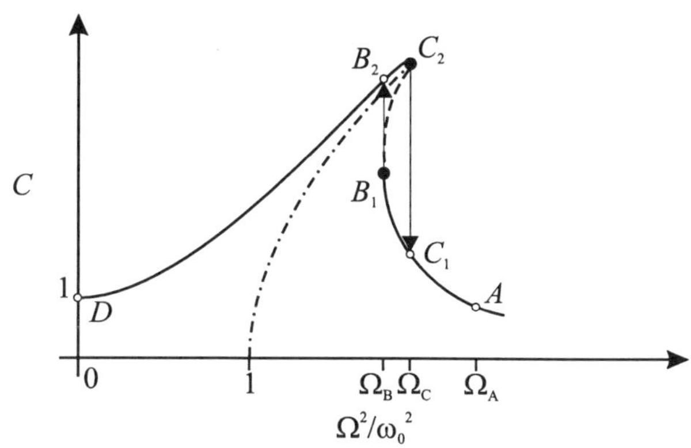


**Figure 73** A typical resonance curve

Now consider the typical resonance curve shown in Figure 73 for a function $f$ with a so called hardening characteristic. Beginning with $\Omega_A$ if one slowly decreases the excitation frequency $\Omega$, then the corresponding point on the resonance curve moves from $A$ to $B_1$. If $\Omega$ is reduced still further from $\Omega_B$, then the system suddenly encounters strong non-stationary (transient) oscillations which jump in magnitude, settling on to a stationary oscillation corresponding to the point $B_2$. For any further decrease in $\Omega$ the representative point on the resonance curve travels from $B_2$ in the direction of $D$. For an increase of the $\Omega$ from 0, the point on the resonance curve moves from $D$ to $C_2$, where it then jumps to $C_1$, as soon as $\Omega_C$ is exceeded. The sudden transitions, $B_1 \rightarrow B_2$ and $C_2 \rightarrow C_1$, are called jump phenomena.

A jump phenomenon is a simple example of crisis (*see* **crisis**) or saddle–node bifurcation, while the process of different patterns of response for increasing and decreasing parameters is called hysteresis.

# K

## KAM (Kolmogorov–Arnold–Moser) theorem

Consider a near-integrable system with a Hamiltonian

$$H = H_0(I) + \epsilon H_1(I, \theta) \tag{1}$$

where $I, \theta \in \mathcal{R}^n$ and

$$\frac{\mathrm{d}I_k}{\mathrm{d}t} = -\frac{\partial H}{\partial \theta_k}$$

and

$$\frac{\mathrm{d}\theta}{\mathrm{d}t} = \frac{\partial H}{\partial I_k}$$

and $k = 1, ..., n$. Near-integrable means that $\epsilon$ is sufficiently small for the following results to hold. Here $H_1$ is periodic in $\theta_1, ..., \theta_N$, and

$$\int_0^{2\pi} ... \int H_1(I, \theta)\mathrm{d}\theta_1 ... d\theta_n = 0,$$

so that

$$H_1(I, \theta) = \sum_{k=1}^{N} h_k(I) e^{ik\theta}. \tag{2}$$

We now look for a generating function, $S(I', \theta)$, of the canonical transformation

$$S = I'\theta + \epsilon S_1 + \epsilon^2 S_2 + ... \tag{3}$$

in which

$$I = I' + \epsilon \frac{\partial S_1}{\partial \theta} + ...$$

and

$$\theta' = \theta + \epsilon \frac{\partial S_1}{\partial I'} + ... \tag{4}$$

transforming the Hamiltonian to a function of only the new actions $I'$. We obtain

$$H_0(I) = H_0(I') + \epsilon \frac{\partial H_0}{\partial I'} \frac{\partial S_1}{\partial \theta} + \epsilon^2 \frac{\partial H_0}{\partial I'} \frac{\partial S_2}{\partial \theta} + O(\epsilon^3) \tag{5}$$

and

$$\epsilon H_1(I,\theta) = \epsilon H_1(I',\theta) + \epsilon^2 \frac{\partial H_1}{\partial I'}\frac{\partial S_1}{\partial \theta} + O(\epsilon^3). \tag{6}$$

Adding (5) and (6), the term proportional to $\epsilon$ can be made to vanish provided that

$$\omega(I')\frac{\partial S_1(I',\theta)}{\partial \theta} = -H_1(I',\theta). \tag{7}$$

If $H_1$ is analytic in a strip about the real $\theta$ axes and periodic in $\theta$, then $H_1(I,\theta)$ is a quasi-periodic function. Also (7) represents a quasi-periodic function for

$$S_1 = \sum_k A_k(I')e^{ik\theta} \tag{8}$$

provided that, for all integer component vectors $k$

$$|k\omega| > C^{-1}|k|^{\mu}. \tag{9}$$

In this case

$$A_k(I') = \frac{ih_k(I')}{k\omega(I')}$$

where $\omega(I) = \partial H_0(I)/\partial I$. As $S_1$ is quasi-periodic, the series in (8) converges. The subsequent iterations for $S_2, S_3, ...,$ etc. will similarly all give quasi-periodic functions. The frequencies $\omega$ can be chosen to satisfy (9) providing that they are independent functions of $I$, i.e.

$$\left|\frac{\partial^2 H_0(I)}{\partial I_i \partial I_k}\right| \neq 0. \tag{10}$$

If (10) holds, then most of the tori will fulfill (9) (with $\mu = N+1$, with the measure going to 1 as $\epsilon \to 0$). However, resonant tori are densely scattered in the phase space, for which (9) is not satisfied, and for which the series (3) diverges. The KAM theorem shows that for most of the tori which fulfill (9), the infinite series of quasi-periodic functions (3) converges for small $\epsilon$.

Let $Q$ be an open set of $\mathcal{R}^n$ and let $H(I,\theta,\epsilon)$ be a real analytic function for all $I \in Q$, $0 \leq \theta \leq 2\pi$, and $\epsilon$ near $\epsilon = 0$. Assume also that $H(I,\theta,\epsilon)$ has period $2\pi$ in each $\theta_1, \theta_2, ..., \theta_N$, and that

$$H_0(I) = H(I,\theta,0)$$

is independent of $\theta$. It is further assumed that for $I \in Q$, equation (10) is satisfied and that the corresponding frequencies $\omega(I) = \partial H_0/\partial I$ fulfill the condition

$$|k\omega| \geq C|k|^{-N}$$

for all integer vectors $|k| = |k_1| + ... + |k_N| \geq 1$. Then, for sufficiently small $\epsilon$ there exist solutions of (1) on invariant tori, defined by

$$\theta' = \theta + F(\theta, \epsilon)$$

and

$$I' = I + G(\theta, \epsilon)$$

where $F$ and $G$ are real analytic functions of $\epsilon$, and $\theta$, having period $2\pi$ in each $\theta_1, ..., \theta_N$, and which vanish for $\epsilon = 0$. Moreover, on these tori the flow satisfies

$$\frac{d\theta'_k}{dt} = \omega_k = \frac{\partial H_0}{\partial I_k}$$

for $k = 1, 2, ..., N$. Finally, the measure of the states on the energy surface which lie on such invariant tori approaches 1 as $\epsilon \to 0$, i.e. for sufficiently small $\epsilon$, the measure of the states on invariant tori is large.

*References*: Arnold (1983), Jackson (1990).

**Kaplan-Yorke conjecture** (*see* **Lyapunov dimension**)

**KdV (Korteweg–de Vries) equation**

The partial differential equation

$$\frac{\partial \phi(x,t)}{\partial t} + (1 + \phi(x,t))\frac{\partial \phi(x,t)}{\partial x} + \frac{\partial^3 \phi(x,t)}{\partial x^3} = 0$$

is called the Korteweg–de Vries or KdV equation.

The KdV equation is one of the fundamental equations in the theory of solitons.

*Reference*: Drazin and Johnson (1989).

**Kermack–McKendrick model**

This model is used to describe the evolution of an epidemic and is given by

$$\frac{dx}{dt} = -kxy$$

$$\frac{dy}{dt} = kxy - ly$$

and

$$\frac{dz}{dt} = ly$$

where $x$ ia the number of susceptible (S) people, $y$ is the number of infected (I) people, $z$ is the number of removed (R) people, $l$ is the rate at which sick people die, and $k$ is the rate at which healthy people become sick. The abbereviation, the SIR model, is sometimes used.

*Reference*: Strogatz (1994).

**kicked rotor**
This is a rotor with rotary inertia, as shown in Figure 74. A bar, of inertial moment $I$ and length $l$, is fastened at one end to a frictionless pivot. At the other end, the bar is subjected to a vertical periodic impulsive force $K/l$, applied at times $t = 0, \tau, 2\tau, ...$, etc. In addition, it is assumed that there is no gravity.

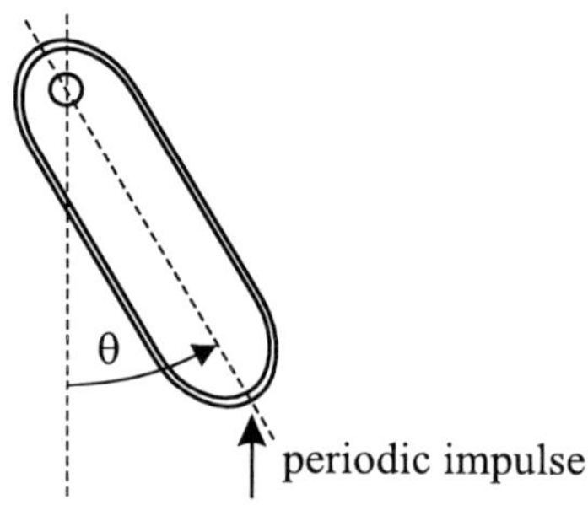


**Figure 74** A kicked rotor

The kicked rotor is a Hamiltonian system for which, unlike most dynamical systems, an analytic expression is derivable for the Poincaré map. The map reduces to

$$\theta_{n+1} = (\theta_n + p_n) \bmod (2\pi)$$

and

$$p_{n+1} = p_n + K \sin \theta_{n+1}$$

where $p_n$ and $\theta_n$ denotes, respectively, the values of the angular momentum and angular position of the rotor at times $t = 0, \tau, 2\tau, ...$, etc., and it is assumed that $\tau/I = 1$.

This map is also called the standard map and has proven to be a very convenient model for the study of typical chaotic behaviour in Hamiltonian systems.

*Reference*: Ott (1992).

**Klein bottle**

This bottle is the simplest example of a non-orientable manifold without a boundary. Consider the diffeomorphism $f$ which maps a circle $S^1$ into itself, i.e. $f : S^1 \to S^1$. Then construct a manifold $\mathcal{M}$, which is the product space of $S^1$ and the unit interval $I = (0, 1)$:

$$\mathcal{M} = S^1 \times I$$

and which satisfies the relationship $(x, 0) = (f(x), 1)$.

If $f(x)$ is orientation-preserving, then $\mathcal{M}$ is a two-torus $T^2$.

If $f(x)$ is orientation-reversing, then $\mathcal{M}$ is a Klein bottle, as shown in Figure 75.

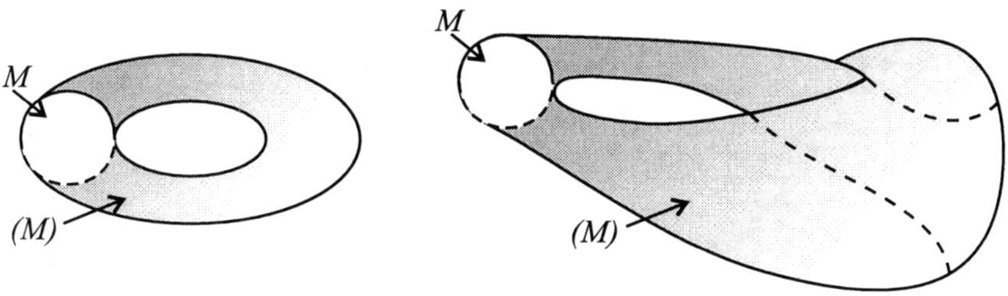


**Figure 75** A Klein bottle

*Reference*: Nitecki (1971).

**Klein–Gordon equation**

The partial differential equation for studying wave dynamics

$$\frac{\partial^2 u}{\partial x^2} - \frac{\partial^2 u}{\partial t^2} = F(x)$$

where $x \in \mathcal{R}$, is called the Klein–Gordon equation.

The sine–Gordon equation (*see* **sine–Gordon equation**) is a particular form of the Klein–Gordon equation.

**kneading theory**

Let $f : [0, 1] \to [0, 1]$ be a map with negative Schwarzian derivative. Assume that $f(0) = f(1) = 0$, and that $f$ has a single critical point $c$, i.e. where $\mathrm{d}f(c)/\mathrm{d}x = 0$ in the interval $[0, 1]$.

Let $f_\mu : [0,1] \to [0,1]$ be a one-parameter family of mappings with $\mu \in [\mu_0, \mu_1]$, which satisfies the above hypothesis and

$$f_{\mu_0}(c) = c$$

and

$$f_{\mu_1}(c) = 1.$$

Such a family is called a full family.

Describing the way in which periodic orbits appear in a full family by using symbolic dynamics is called a kneading theory. It describes most of the structure of an individual map and determines much of the bifurcation structure in the family.

A symbolic description of a map $f$ can be phrased in terms of the partition of the interval $[0,1]$ into the two sub-intervals, $I_0 = [0,c)$, and $I_1 = (c,1]$, and the point $C = \{c\}$. The intervals $I_0$ and $I_1$ are the laps of $f$. If $x \in I_0$, the $n$th address $A_n(x), n \geq 0$ is $I_0, C$ or $I_1$ as $f^n(x) \in I_0$, $f^n(x) = c$, or $f^n(x) \in I_1$. The itinerary of $x$, $A(x)$, is the sequence $\{A_n(x)\}$ of its successive addresses. The itinerary of $f(c)$ is called the kneading sequence.

*References*: Guckenheimer and Holmes (1983); Milnor and Thurston (1977).

**knots**
A knot is a simple closed curve in a $\mathcal{R}^3$.

A collection of periodic orbits of a three-dimensional flow forms a disjoint union of knots, or a braid (link).

Two knots are said to be equivalent if they are topologically equivalent (*see* **topological equivalence**).

*Example 1*: There exists an open set of parameters $\beta \in [6.5, 10.5]$ for which periodic solutions to the differential equations

$$\frac{dx}{dt} = 7[y - \phi(x)]$$

$$\frac{dy}{dt} = x - y + z$$

and

$$\frac{dz}{dt} = -\beta y$$

where

$$\phi(x) = \frac{2}{7}x - \frac{3}{14}[|x+1| - |x-1|]$$

contain representatives from every knot and link equivalence class.

A template or knot-holder is a compact branched two-manifold fitted with a smooth expansive semiflow and built from a finite number of joining and splitting charts.

*Example 2*: The simplest non-trivial template is the Lorenz template shown in Figure 76.

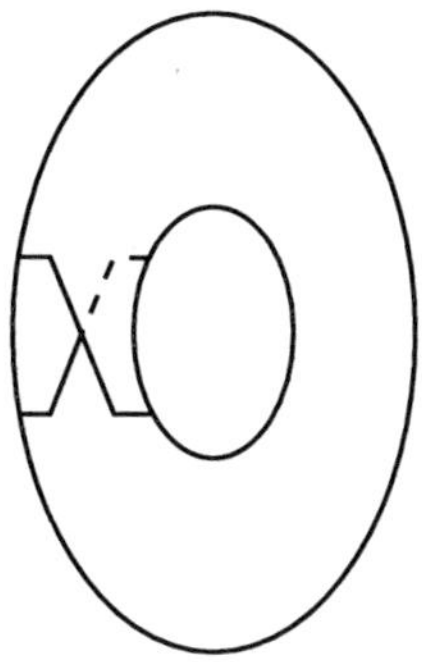

**Figure 76** The Lorenz template

*Reference*: Ghrist and Holmes (1996).

**Koch curve**
Consider the construction of the set shown in Figure 77 which is used as the generator (each segment length is 1/3 on the unit interval). The generator is then used on each of these straight segments and the process is continued infinitely to form a Koch curve.

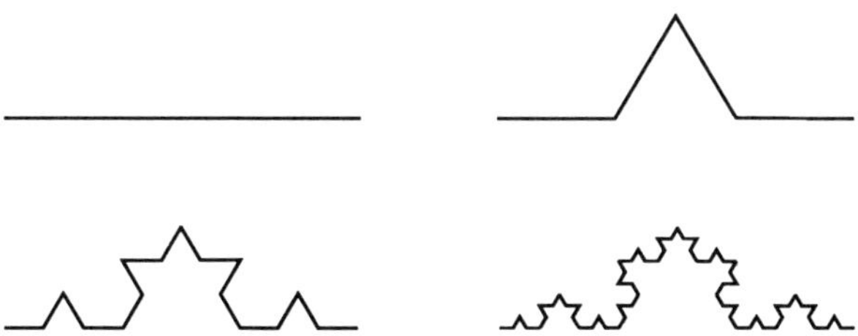

**Figure 77** First steps in the construction of a Koch curve

**Kolmogorov–Arnold–Moser theorem** (*see* **KAM theorem**)

**Kolmogorov capacity dimension** (*see* **dimension of sets**)

**Krylov-Bogoliubov method** (*see* **averaging methods**)

**K-set**
The catastrophe set (*see* **Thom's theorem**) is called a $K$-set.

The catastrophe $K$-set is a set of bifurcation points.

*Reference*: Thom (1975).

# L

**Laffer curve**
This is the curve that relates the taxation rate to the government's revenue. This curve is discussed in connection with the application of nonlinear dynamics to economy.

*Reference*: Jackson (1990).

**Lagrange stability** (*see* **stability**)

**lambda ($\lambda$) lemma**
Let $f$ be a diffeomorphism of the plane $f : \mathcal{R}^2 \to \mathcal{R}^2$ and let $p$ be a hyperbolic saddle (fixed point) of $f$. Suppose that a curve $L$ crosses the stable manifold of $p$ transversally. Then each point in the unstable manifold of $p$ is a limit point of $\cup_{n>0} f^n(L)$.

This lemma is used to explain sudden changes in the appearance of chaotic attractors as parameters vary.

*References*: Alligood *et al.* (1997); Palis and de Melo (1982).

**Landau equation**
An ordinary differential equation

$$\frac{\mathrm{d}x}{\mathrm{d}t} = ax - bx^3$$

where $x \in \mathcal{R}$, and $a$, and $b$ are constants, is called the Landau equation.

This equation was used by Landau to investigate the stability of the steady flow of a Newtonian fluid.

*Reference*: Drazin (1992).

**Landau–Hopf theory**
The classical theory of turbulence which, at the onset of turbulence, generates a countably infinite number of frequencies (after a countably infinite number of Hopf bifurcations) is called the Landau–Hopf theory. This may be contrasted to the Ruelle–Takens approach to turbulence.

**Laplace's equation**
The partial differential equation given by

$$\frac{\partial^2 u}{\partial x^2} + \frac{\partial^2 u}{\partial y^2} = 0$$

where $x, y \in R$, is called Laplace's equation.

The equation can be extended to any dimension $n$.

This equation was initially developed by Laplace and is now used for problems involving steady-state heat flow and vibrating shells, among other things.

*Reference*: Smith (1969).

**lattice maps** (*see* **cellular automata** and **coupled map lattice**)

**Lebesgue measure**
This measure represents the 'length' of a set generalised to $\mathcal{R}^n$ (or any Riemannian manifold).

Let $S \subset \mathcal{R}$ be a bounded non-empty open set. It can be represented as the union of a finite or a countably infinite number of disjoint open intervals whose end points do not belong to $S$, i.e.

$$S = \cup_k (a_k, b_k).$$

The Lebesgue measure of the open set $S$ is

$$\mu(S) = \sum_k (b_k - a_k).$$

Having a Lebesgue measure $\mu$ in the measurable space $(\mathcal{R}, \mathcal{B}, \mu)$ we can generalise them to $\mathcal{R}^n$ by defining the product space $\mathcal{R}^n, \mathcal{B}^n$, and $\mu^n$, where

$$\mathcal{R}^n = \mathcal{R} \times \mathcal{R} \times ... \times \mathcal{R}$$

with $\mathcal{B}^n$ being the smallest $\sigma$-algebra containing all sets of the form $S_1 \times ... \times S_n$ with $S_i \in \mathcal{B}$ and

$$\mu^n(S_1 \times ... \times S_n) = \mu(A_1)....\mu(A_n).$$

A non-empty closed set $S \subset \mathcal{R}$ is either a closed interval or else can be obtained from a closed interval by removing a finite or countably infinite family of disjoint open intervals whose end points belong to $S$. Thus a closed set $S$ can be expressed as

$$S = [a, b] - \sum_k (a_k - b_k).$$

The Lebesgue measure of the closed set $S$ is

$$\mu(S) = (b - a) - \sum_k (b_k - a_k).$$

*References*: Edgar (1990), Ott (1992).

**Liapunov** (*see* **Lyapunov**)

**Liénard's method**

This is a graphical method of solving second-order differential equations of the special type

$$\frac{\mathrm{d}^2 x}{\mathrm{d}t^2} + f\left(\frac{\mathrm{d}x}{\mathrm{d}t}\right) + x = 0. \tag{1}$$

By putting $y = \mathrm{d}x/\mathrm{d}t$, equation (1) becomes

$$\frac{\mathrm{d}y}{\mathrm{d}x} = -\frac{x + f(y)}{y}.$$

Assuming that $x(0) = x_0$ and $\mathrm{d}x(0)/\mathrm{d}t = y_0$, then the initial point in the phase space is $(x_0, y_0)$. Liénard's method starts by drawing the curve $x + f(y) = 0$ in the phase space, as shown in Figure 78.

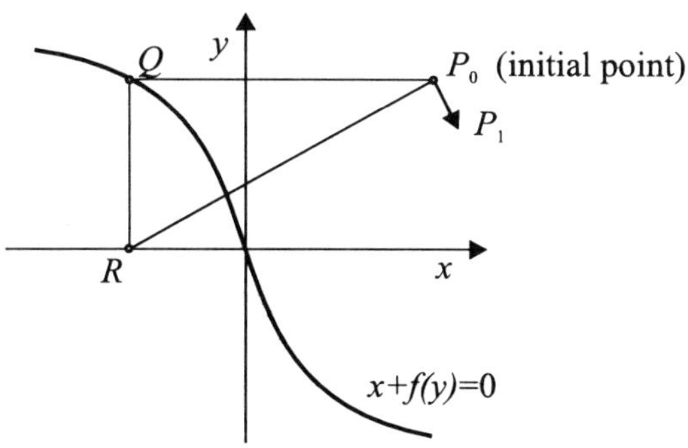


**Figure 78** Illustration of Liénard's method

A line parallel to the $x$-axis is drawn from the initial point $P_0(x_0, y_0)$, to intersect $x + f(y) = 0$ at $Q$. The line $QR$ is perpendicular to the $x$-axis at $R$. It is easily shown that the gradient of the line $RP_0$ is

$$\frac{y_0}{x_0 + f(y_0)} = -\frac{1}{(\mathrm{d}y/\mathrm{d}x)_0}.$$

Therefore, the integral curve though $P_0$ is perpendicular to $RP_0$. A small segment $P_0P_1$ is drawn perpendicular to $RP_0$ and the process is repeated starting from $P_1$.

*Reference*: Huntley and Johnson (1983).

**limit cycle**
A periodic orbit $A$ of the autonomous system

$$\frac{\mathrm{d}x}{\mathrm{d}t} = f(x)$$

where $x \in \mathcal{R}^n$, is called a limit cycle, where the latter may be stable or unstable. If $A$ is asymptotically stable, then the limit cycle is stable. If $A$ is not asymptotically stable, then the limit cycle is unstable.

**limit point**
The point $x$ is a limit point of a set $A$ if every neighbourhood of $x$ contains a point in $A - \{x\}$.

A point $q$ is an $\omega(\gamma)$-limit point of $x_0$ if there exists a sequence of times $\{t_k\}$, where $t_k \to \infty$ as $k \to \infty$, and such that $q$ is the limit of the orbit $\gamma(x_0)$ in the future, i.e.

$$\lim_{k\to\infty} x(t_k, x_0) = q, \qquad t_k \to \infty.$$

A point $q$ is an $\alpha(\gamma)$-limit point of $x_0$ if there exists a sequence of times $\{t_k\}$, where $t_k \to -\infty$ as $k \to \infty$, and such that $q$ is the limit of the orbit $\gamma(x_0)$ in the past, i.e.

$$\lim_{k\to\infty} x(t_k, x_0) = q, \qquad t_k \to -\infty.$$

*Reference*: Jackson (1990).

**Lindstedt–Poincaré method**
This method is a perturbation method for the solution of nonlinear ordinary differential equations producing uniformly valid expansions. Typically, this method is an improvement on the straightforward expansion method, allowing

flexibility to 'remove' non-uniform terms as they appear, and which uses a frequency $\omega$, dependent upon a dimensionless time $\tau = \omega t$.

*Example*: Consider the equation

$$\frac{\mathrm{d}^2 u}{\mathrm{d}t^2} + u + \epsilon u^3 = 0$$

where $\epsilon \ll 1$. With the change of variable $\tau = \omega t$, this equation becomes

$$\omega^2 \frac{\mathrm{d}^2 u}{\mathrm{d}\tau^2} + u + \epsilon u^3 = 0.$$

The solution is approximated by

$$u(\tau, \epsilon) = u_0(\tau) + \epsilon u_1(\tau) + \epsilon^2 u_2(\tau) + ...$$

and

$$\omega = 1 + \epsilon\omega_1 + \epsilon^2 \omega_2 + ...$$

where $\omega_n$ are constants to be determined from the condition corresponding to the vanishing of secular terms. Then $u$ and $\omega$ are substituted into the differential equation and terms including powers of $\epsilon$ are collected together (*see also* **averaging methods**).

*Reference*: Nayfeh (1985).

**linearisation**
Let $u^*$ be a fixed point of $\mathrm{d}u/\mathrm{d}t = f(u)$ $u \in \mathcal{R}^n$, i.e.

$$f(u^*) = 0.$$

In analyzing fixed points, we linearise the differential equation in a neighbourhood of the fixed point. Let us assume that $f$ is analytic. Thus we have a Taylor-series expansion of $f$ around $u^*$. Linearising means that we neglect second- and higher-order terms. In the case of

$$\frac{\mathrm{d}u}{\mathrm{d}t} = f(u)$$

we can write in the neighbourhood of the fixed point $u^*$

$$\frac{\mathrm{d}u}{\mathrm{d}t} = \frac{\partial f}{\partial u}(u - u^*) + \text{ higher-order terms}$$

and study the linear differential equation

$$\frac{\mathrm{d}u}{\mathrm{d}t} = \frac{\partial f}{\partial u}(u^*)(u - u^*).$$

The $n \times n$ matrix $\partial f/\partial u$ is called the *Jacobian matrix*. In order to simplify the notation, the fixed point $u^*$ is shifted to the origin of the phase space by $\bar{y} = u - u^*$. Thus

$$\frac{\mathrm{d}\bar{y}}{\mathrm{d}t} = \frac{\partial f}{\partial u}(u^*)\bar{y}.$$

We often write $\partial f(u^*)/\partial u$ as $A$, i.e. a $(n\times n)$-dimensional matrix with constant coefficients. Omitting the bar, the linearised system in the neighbourhood of a fixed point $u^*$ is of the form

$$\frac{\mathrm{d}y}{\mathrm{d}t} = Ay \qquad (1)$$

and if the linearisation is hyperbolic then we can analyse the linear dynamics.

Linearised equations of the form (1) are also sometimes called variational equations, essentially when linearised about a periodic solution.

*Example*: Consider the mathematical pendulum

$$\frac{\mathrm{d}^2u}{\mathrm{d}t^2} + \sin u = 0$$

where $u$ is the angular variable indicating the deviation from the vertical, so $u \in (-\pi,\ \pi)$. After setting $u = u_1$, and $\mathrm{d}u/\mathrm{d}t = u_2$ one obtains the following:

$$\frac{\mathrm{d}u_1}{\mathrm{d}t} = u_2, \qquad \frac{\mathrm{d}u_2}{\mathrm{d}t} = -\sin u_1.$$

Linerisation in the neighbourhood of $(0,0)$ yields

$$\frac{\mathrm{d}u_1}{\mathrm{d}t} = u_2, \qquad \frac{\mathrm{d}u_2}{\mathrm{d}t} = -u_1.$$

Linearisation in the neighbourhood of $(\pm\pi,0)$ gives

$$\frac{\mathrm{d}u_1}{\mathrm{d}t} = u_2, \qquad \frac{\mathrm{d}u_2}{\mathrm{d}t} = u_1 \mp \pi.$$

**Liouville theorem**

Let $D(0)$ represent an arbitrary domain in the phase space at $t = 0$, and let $D(t)$ represent this domain at the time $t$; therefore $D(t)$ is the set of all points $x(t, x^0)$ which evolve according to the relationship

$$\frac{\mathrm{d}x}{\mathrm{d}t} = f(x,c,t) \qquad (1)$$

where $x \in \mathcal{R}^n$, and $c \in \mathcal{R}^m$, and such that $x(0) = x^0$ is contained in $D(0)$. Consider the integral of some function $\rho(x,t) : \mathcal{R}^n \times \mathcal{R} \to \mathcal{R}$ over this moving domain $D(t)$:

$$I(t) = \int ... \int_{D(t)} \rho(x,t)\mathrm{d}x_1 ... \mathrm{d}x_n.$$

The theorem states that the integral $I(t)$ is an integral invariant of equation (1), if and only if

$$\frac{\partial \rho}{\partial t} + \nabla(\rho(x,t)f(x,t)) = 0. \tag{2}$$

$I(t)$ is called an integral invariant of (1).

Equation (2) is called Liouville's equation.

*Remark*: When $\rho(x,t) = 1$ for all $x \in \mathcal{R}^n$, and $t \in \mathcal{R}$ then the phase volume is preserved.

*Reference*: Jackson (1990).

**Lipschitz condition**
The function $f(x) : D \to \mathcal{R}^n$ fulfills the Lipschitz condition if there exists a constant $K < \infty$ such that

$$|f(y) - f(x)| \leq K|y - x|$$

for all $x, y \in D$.

**local**
A property is local if it can be analysed in an arbitrarily small neighbourhood of a given point.

**Li–Yorke chaos**
The term chaos was first used by T Y Li and J A Yorke in a paper entitled 'Period three implies chaos'. This paper examined the Sarkovskii theorem and proposed a method to prove the existence of an infinite number of solutions for a particular class of maps (*see also* **Sarkovskii theorem**). The idea of Li–Yorke chaos is related to a topological horseshoe (*see* **horseshoe map**).

The map $f : I \to I$, where $I$ is a closed interval, is called chaotic in the sense of Li–Yorke if:

(i) $f$ has points $x \in I$ of an arbitrary long period;

(ii) there is an uncountable set $S$ ($S \subset I$) consisting of non-periodic points which only satisfying the following conditions:

(a) for every $x, y \in S$ with $x \neq y$

$$\lim_{n\to\infty} \sup |f^n(x) - f^n(y)| > 0$$

and

$$\lim_{n\to\infty} \inf |f^n(x) - f^n(y)| > 0$$

(b) for every $x \in S$ and every periodic point $y \in I$

$$\lim_{n\to\infty} \sup |f^n(x) - f^n(y)| > 0.$$

*Reference*: Li and Yorke (1975).

**lobe**

This is a term used in geometric investigations of a homoclinic (or heteroclinic) tangle, i.e. the intersection of the stable $W^s$ and unstable $W^u$ manifolds of a saddle point $S$. The lobe is a portion of the phase space trapped between two consecutive crossings of the manifolds, as shown in Figure 79.

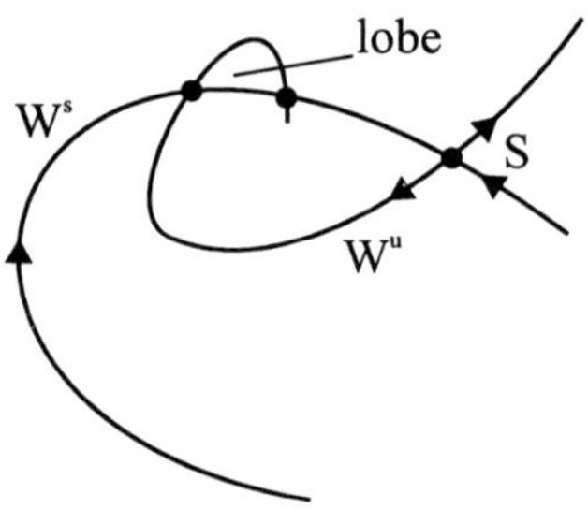


**Figure 79** Illustration of a lobe

*Reference*: Wiggins (1990).

**local bifurcations**

Bifurcations which occur in the neighbourhood of fixed points or periodic orbits are referred to as local bifurcations.

**logistic equation**

The first-order ordinary differential equation

$$\frac{du}{dt} = cu(1-u) \tag{1}$$

where $c \geq 0$, is called the logistic equation.

The logistic equation was formulated by the Belgian social statistician Pierre-Francois Verhulst as an extension of the Malthusian model of growth of a population. The equation is usually given in the form

$$\frac{\mathrm{d}N(t)}{\mathrm{d}t} = rN(t)\left\{1 - \frac{N(t)}{N(\infty)}\right\}$$

where $N(t)$ is the population at time $t$, $r$ is a constant which represents fecundity, and $N(\infty)$ the maximal population size (*see also* **Malthusian growth**).

In population dynamics, it is often assumed that the reproduction is episodic rather than continuous in time, so that for example the population each year can be expressed as a function of the previous year, thus leading to a discrete form of the equation

$$x_{n+1} = rx_n(1 - x_n)$$

which has very different dynamics to the flow defined by equation (1).

**logistic map**
The quadratic map

$$x_{n+1} = ax_n(1 - x_n) \tag{1}$$

is called the logistic map.

If $1 < a < 3$, the fixed point at $x^* = 1 - 1/a$ is an attractor (as shown in Figure 80(a)), and the system settles to the stable point. At $a = 3$, the system bifurcates, to give a cycle of period 2 (Figure 80(b)), which is stable for $3 < a < (1 + \sqrt{6})$. As $a$ increases beyond this, successive bifurcations give rise to a cascade of period-doublings, producing cycles of periods 4 (Figure 80(c)), 8, 16, ..., $2^n$. With a further increase of $a$, we observe an apparently chaotic régime, in which trajectories look like sample functions of random processes (Figure 80(d)).

On a fine scale, the chaotic régime is in fact comprised of infinitely many windows of $a$-values in which basic cycles of period $k$ are born stable (accompanied by unstable twins). These undergo cascades of period-doublings to give stable harmonics of period $k \times 2^n$, and become unstable; this sequence of events recapitulates the process seen more clearly for the basic fixed point of period 1. The bifurcation diagram for $2.9 < a < 4$ is shown in Figure 81; for $a > 4$ all trajectories diverge.

The nature of the chaotic régime for such 'maps of the interval' is often misunderstood. In detail, the chaotic régime is largely a mosaic of stable cycles,

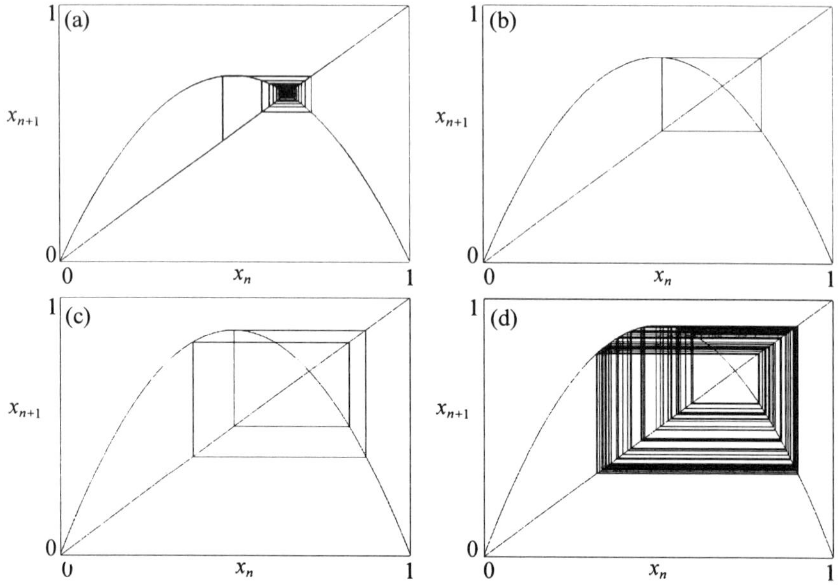


**Figure 80** Different types of attractors of a logistic map – equation (1): (a) fixed point; (b) period 2; (c) period 4; (d) chaos

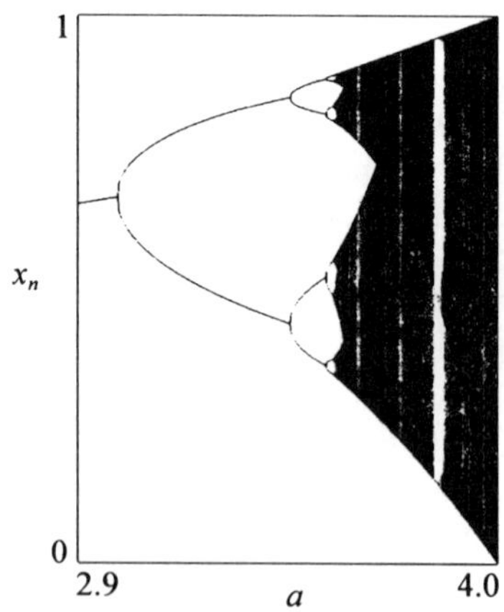


**Figure 81** Bifurcation diagram of the logistic map

with one giving way to another with great rapidity as $a$ increases. This point is exemplified by the Lyapunov exponent which is often computed as an index of chaotic behaviour. These exponents are analogous to the eigenvalues that characterise the stability properties of simpler systems. They are typically calculated by iterating difference equations, and calculating the geometric average value of the slope of the map at each iterate, i.e. for the difference equation

$$x_{n+1} = f(x_n)$$

the Lyapunov exponent $\lambda$ is given by

$$\lambda = \lim_{n\to\infty} \left\{ \frac{1}{n} \sum_{i=0}^{n-1} \ln \frac{\mathrm{d}f(x_n)}{\mathrm{d}x} \right\}. \tag{2}$$

For generic quadratic maps, there are unique attractors for most values of $a$ in the chaotic régime. Therefore in calculation (2), if carried out exactly, or if the iterations are carried out for long enough, $\lambda$ is typically found to be positive in the chaotic régime. A plot of the Lyapunov exponent for the Logistic map is shown in Figure 82.

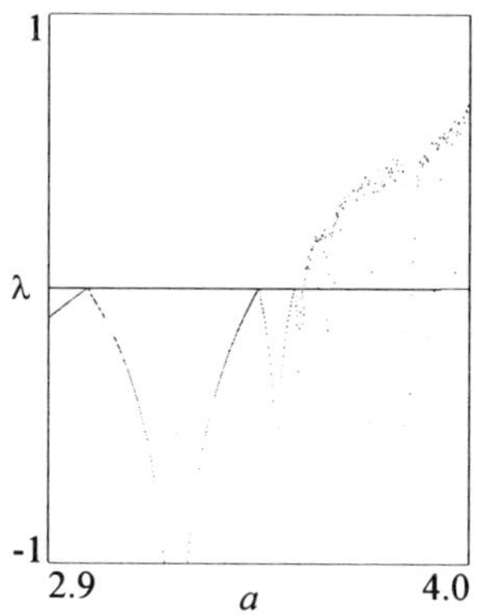


**Figure 82** Lyapunov exponent of the logistic map

*References*: Feigenbaum (1978); Ott (1992).

**Lorenz model**

In 1963, E. Lorenz proposed a three-dimensional model for the atmospheric convective flow produced by truncation of the Navier–Stokes equation. The dynamical system given by the equations

$$\frac{\mathrm{d}x}{\mathrm{d}t} = -\sigma(x - y)$$

$$\frac{dy}{dt} = -xz + rx - y$$

and

$$\frac{dz}{dt} = xy - bz \tag{1}$$

where $\sigma, r$ and $b$ are dimensionless parameters, is called the Lorenz system.

*Example*: For $\delta = 10, b = 8/3$ and $r = 28$, the Lorenz system shows chaotic behaviour.

Figure 83 shows a projection of the phase-space trajectory on to the $xz$ plane. The points $C_1$ and $C_2$ represent projections of fixed points which are unstable for the considered parameter values. It can be observed that the solution spirals outward from one of the fixed points $C_1$ or $C_2$ for some time, then switches to spiraling outward from the other fixed point. This pattern repeats forever with the number of revolutions around a fixed point before switching, varying in an apparently random manner.

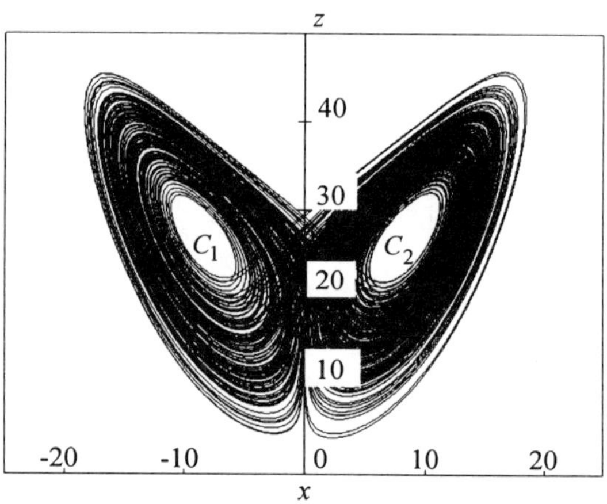


**Figure 83** Phase-space trajectory of the Lorenz system

Let $m_n$ be the $n$-th maximum of the function $z(t)$. If we plot $m_{n+1}$ versus $m_n$, we obtain a one-dimensional map as shown in Figure 84, whose dynamic is very similar to the so-called tent map:

$$x_{t+1} = 1 - 2\left|\frac{1}{2} - x_t\right|.$$

*References*: Lorenz (1963); Sparrow (1982).

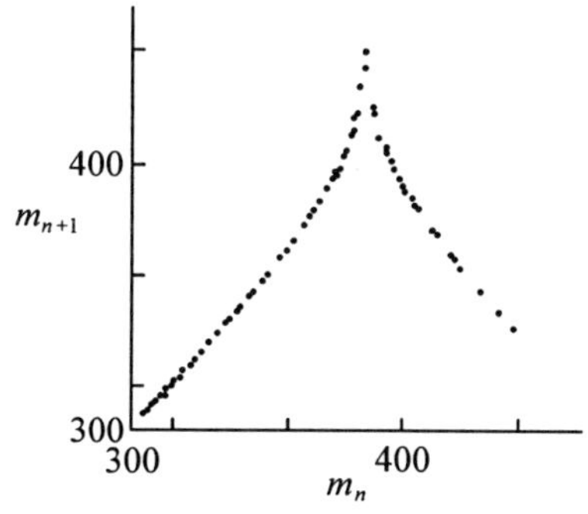


**Figure 84** One dimensional map of $m_{n+1}$ versus $m_n$

**Lotka–Volterra equations**
These equations, given by

$$\frac{du_1}{dt} = au_1 - bu_1u_2$$

and

$$\frac{du_2}{dt} = bu_1u_2 - cu_2$$

with $u_1, u_2 \geq 0$, and where $a, b, c$ are positive constants, describe the interaction of two species, where $u_1$ denotes the population density of the prey and $u_2$ the population density of the predator.

*Reference*: Ebeling and Peschel (1985).

**low-dimensional systems**
These systems are typically taken to mean flows of dimension $d \leq 3$, or maps of dimension $d \leq 2$.

**Lozi map**
The two-dimensional real map

$$x_{n+1} = 1 + y_n - a|x_n|$$

and

$$y_{n+1} = bx_n$$

where $x, y \in \mathcal{R}$, $a, b \in \mathcal{R}$, $a, b > 0$ is called the Lozi map.

*Remark*: The Lozi map is a simplification, or piecewise linear version, of the Hénon map.

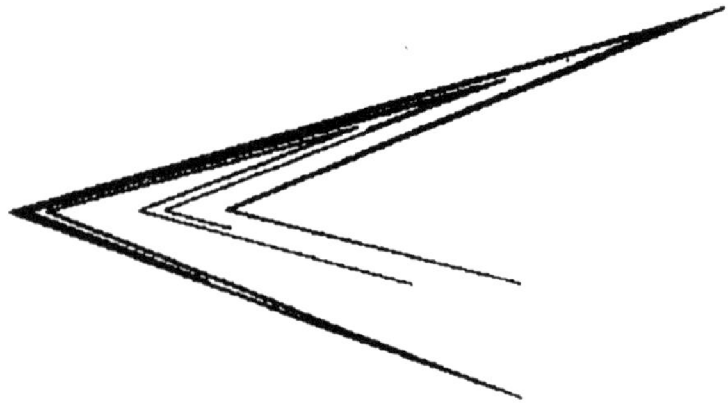

**Figure 85** Chaotic behaviour of Lozi map in the $(x, y)$ plane

*Example*: For $a = 1.7$ and $b = 0.5$, there exists the strange attractor shown in Figure 85.

**$L^p$-space**

Let $(X, \mathcal{B}, \mu)$ be a measure space and $p$ a real number $1 \leq p < \infty$. The family of all possible real-valued measurable functions $f : X \to R$ satisfying

$$\int_X |f(x)|^p \, \mu(\mathrm{d}x) < \infty$$

is the $L^p(X, \mathcal{B}, \mu)$ space.

*Reference*: Lasota and Mackey (1985).

**Lyapunov dimension**

The dimension of a fractal chaotic attractor can be associated with the corresponding Lyapunov exponents. Let a typical trajectory on an attractor be characterised by the following Lyapunov exponents

$$\lambda_1, \lambda_2, ..., \lambda_n \ (\lambda_i \geq \lambda_{i+1}, \ i = 1, 2, ..., n-1).$$

The *Lyapunov dimension* $d_L$ is defined by

$$d_L = j + \frac{\sum_{i=1}^{j} \lambda_i}{|\lambda_{j+1}|}$$

where $j$ is the maximum index for which

$$\sum_{i=1}^{j} \lambda_i \geq 0.$$

Kaplan and Yorke shows that for commonly found attractors, the Lyapunov dimension is numerically close to the information dimension.

*Reference*: Kaplan and Yorke (1979).

**Lyapunov exponents**

These exponents, named after the Russian mathematician A.M. Lyapunov, can be used to obtain a measure of the sensitive dependence of initial conditions for a solution of the system

$$\frac{du}{dt} = f(u), \qquad u \in D \subset \mathcal{R}^n.$$

Lyapunov exponents are a generalisation of the eigenvalues of a fixed point. The linearised equation is given by

$$\frac{dy}{dt} = \frac{\partial f}{\partial u}(\Phi(u_0)y)$$

where $u_0 = u(t=0)$. The solution of this system can be written as

$$y(t) = U^t_{u_0} y_0$$

where $U^t_{u_0}$ is the matrix of fundamental solutions. The fundamental matrix satisfies the chain rule

$$U^{t+s}_{u_0} = U^t_{u_0} \circ U^s_{u_0}$$

The asymptotic behaviour of the fundamental matrix for $t \to \infty$ can be characterised by the following exponents

$$\lambda(V^k, u_0) = \lim_{t\to\infty} \ln \frac{U^t_{u_0}\mathbf{e_1} \wedge U^t_{u_0}\mathbf{e_2} \wedge ... \wedge U^t_{u_0}\mathbf{e_k}}{||\mathbf{e}_1 \wedge \mathbf{e}_2 \wedge \ldots \wedge \mathbf{e}_k||},$$

where $\wedge$ indicates the outer product.

Let $m_1(t), \ldots, m_n(t)$ be the eigenvalues of the solution of

$$\frac{d\mathbf{y}}{dt} = A(u_0)\mathbf{y}$$

given by

$$\mathbf{y}(t) = e^{A(u_0)t}$$

where

$$A(x_0) = \frac{\partial \mathbf{f}}{\partial u}(u_0).$$

The Lyapunov exponents of $x_0$ are

$$\lambda_i(u_0) = \lim_{t\to\infty} \frac{1}{t} \ln |m_i(t)|$$

whenever the limit exists. For the existence of Lyapunov exponents, *see* **Oseledec theorem**.

*Example*: Consider the Lyapunov exponents of the fixed point $u^*$. Let $\hat{\lambda}_1, \ldots, \hat{\lambda}_n$, be the eigenvalues of $A(u^*)$; then $m_i(t) = e^{\hat{\lambda}_i t}$, and

$$\lambda_i = \lim_{t\to\infty} \frac{1}{t} \ln |e^{\hat{\lambda}_i t}| = \lim_{t\to\infty} \frac{1}{t} \text{ Re } [\hat{\lambda}_i]t = \text{Re}[\hat{\lambda}_i].$$

As shown in the above example, Lyapunov exponents are equal to the real parts of the eigenvalues of the fixed point. They indicate the rate of contraction (when $\lambda_i < 0$) or expansion (when $\lambda_i > 0$) close to the critical point. The subspaces in which the expansion or contraction occurs are determined by the appropriate eigenvectors of $A(u^*)$.

*Remark*: Positive one-dimensional Lyapunov exponents mean that two nearby trajectories (trajectories for slightly different initial conditions) diverge exponentially.

*References*: Benettin *et al.* (1980); Wolf *et al.* (1985).

**Lyapunov first theorem**
Consider the system

$$\frac{dx}{dt} = f(x)$$

where $x \in \mathcal{R}^n$, for which $x_0$ is a fixed point. This theorem states that $x_0$ is

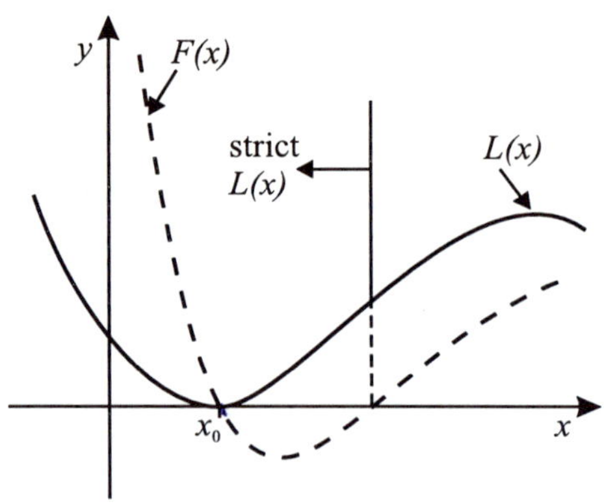


**Figure 86** Representation of the Lyapunov function

a stable fixed point if there exists a function $L(x)$ such that:

(i) $L(x_0) = 0$ and $L(x) > 0$, if $x_0 \neq x$ in some region $\Omega$ about $x_0$;

(ii) $f \cdot \nabla L \leq 0$ in $\Omega$, where $\nabla$ is a gradient operator.

The function $L(x)$ which satisfies these conditions is called a Lyapunov function. An example of such a function is shown in Figure 86.

If $f \cdot \nabla L < 0$, then $L$ is called a strict Lyapunov function.

*Reference*: Jackson (1990).

**Lyapunov function** (*see* **Lyapunov first theorem**)

**Lyapunov stability** (*see* **stability**)

# M

## Mackey–Glass equation

This is the differential-delay equation

$$\frac{\mathrm{d}X}{\mathrm{d}t} = \frac{aX(t-s)}{1+[X(t-s)]^c} - bX(t)$$

where $a, b, c$ and the time delay $s$ are constant.

This equation models neutrophil dynamics. For $a = 0.2$, $b = 0.1$, $c = 10.0$ and $s = 31.8$, the equation shows chaotic behaviour.

*Reference*: Glass and Mackey (1988).

## magneto-elastic mechanical oscillator

A magneto-elastic mechanical oscillator is a physical model consisting of a sinusoidally driven beam whose free end oscillates between two magnets. The model was proposed by Moon and Holmes and is modelled by a Duffing equation.

*Reference*: Guckenheimer and Holmes (1983).

## Malthusian growth

At the end of the eighteenth century, T.R. Malthus discussed the rate of growth of a population size in terms of the present population $N(t)$. In present-day terms, we represent this growth by the equation

$$\frac{\mathrm{d}N(t)}{\mathrm{d}t} = rN(t)$$

where $r$ represents the fecundity of the population (i.e. how good the population is at reproducing). If, at some time $t = 0$, the population has a size $N(t = 0) = N_0$, then at some later time the population will be given by

$$N(t) = N_0 \exp(rt).$$

For $r > 0$ we have exponential growth, for $r < 0$ exponential decay, and for $r = 0$ the population size remains constant. For a general $r$, the only way to achieve a constant population is to have $N(t) = 0$, i.e. no population at all!

*Reference*: Grebogi and Yorke (1997).

**Mandelbrot set**

Let $\hat{\mathcal{C}}$ be the Riemmanian sphere, i.e. $\hat{\mathcal{C}} = \mathcal{C} \cup \{\infty\}$. Consider the family

$$\{\hat{\mathcal{C}} : f_\lambda = z^2 - \lambda\}$$

where the parameter space is $P = \mathcal{C}$, i.e. $\lambda \in \mathcal{C}$. Then, the Mandelbrot set $\mathcal{M}$ for the family of the dynamical systems $\{\hat{\mathcal{C}} : z^2 - \lambda\}$ is defined as

$$\mathcal{M} = \{\lambda \in P : J(\lambda), \text{ is connected}\}.$$

An example of a Mandelbrot set is shown in Figure 87.

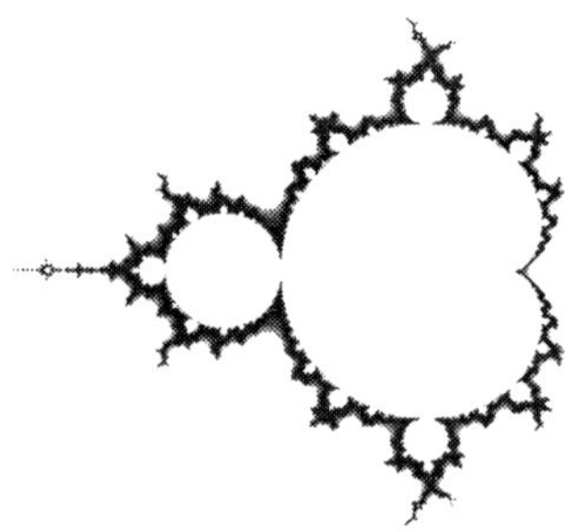

**Figure 87** Illustration of a Mandelbrot set

*Remark*: The Mandelbrot set is not self-similar.

*Reference*: Mandelbrot (1982).

**manifold**

An $n$-dimensional manifold $M \subset \mathcal{R}^n$ is a set for which each $x \in M$ has an open neighbourhood $U$ for which there is a homeomorphism $\phi$, such that for any open set $D \subset \mathcal{R}^n$ we have $\phi : D \to U$ $(n \le N)$.

**map**

A map $f : X \to Y$ associates with each element $x \in X$ one element $y \in Y$, and is denoted by $y = f(x)$.

If each element of $Y$ is associated with some element of $X$, then the mapping is said to be onto, i.e. $f(X) = Y$, or surjective. Otherwise, it is into, i.e. $f(X) \subset Y$.

If no two elements of $X$ are associated with the same element of $Y$, then the map is said to be one-to-one or injective (*see also* **discrete dynamical system**).

**Markov property**
In a deterministic dynamical system $X_{n+1} = f(X_n)$, only the present value $x_n^i$ of $X_n$ is needed to determine the future value of $X_{n+1}$; the past values of $X_1, X_2, ..., X_{n-1}$, are involved indirectly in that they determine the value of $X_n$. This is just the common law of causality for which there is a stochastic analogue called the Markov property, expressed by the conditional probabilities

$$P(X_{n+1} = x^j | X_n = x_n^i) = P(X_{n+1} = x^j | X_1 = x_1^i, X_2 = x_2^i, ..., X_n = x_n^i)$$

for all possible $x_j, x_i^1, ..., x_i^n$, and all $n = 1, 2, ...,$ etc..

A sequence of discrete valued random variables with this property is an example of a Markov chain.

*Reference*: Kloeden and Platen (1992).

**massive bifurcation**
A bifurcation at the transition from regular to chaotic scattering (*see also* **bifurcation** and **chaotic scattering**).

**Mather set** (*see* **Aubry–Mather theorem**)

**Mathieu equation**
This equation, given by

$$\frac{\mathrm{d}^2 x}{\mathrm{d}t^2} + (\alpha + \beta \cos t)x = 0 \tag{1}$$

or the nonlinear version

$$\frac{\mathrm{d}^2 x}{\mathrm{d}t^2} + (\alpha + \beta \cos t)\sin x = 0 \tag{2}$$

where $x \in \mathcal{R}$, and $\alpha$ and $\beta$ are constant, describes a parametrically excited system which physically may be related to the behaviour of a pendulum whose pivot is vertically driven, and is an example of a driven system which can display a state of equilibrium or fixed point.

The linear Mathieu equation (1) is a particular form of Hill's equation.

**Maxwell–Bloch equations**

These equations are a model for laser dynamics, describing the dependence of an electric field $E$, the mean polarisation $P$ of the atoms and the amount of population inversion $D$:

$$\frac{\mathrm{d}E}{\mathrm{d}t} = -kE + kP$$

$$\frac{\mathrm{d}P}{\mathrm{d}t} = Y_1 ED - Y_1 P$$

and

$$\frac{\mathrm{d}D}{\mathrm{d}t} = Y_2(\lambda + 1) - Y_2 D$$

where the decay rate in the laser cavity is given by $k$, the decay rate of the atomic polarisation is given by $Y_1$, the decay rate of the population inversion is given by $Y_2$, and $\lambda$ is the pumping energy parameter.

These equations can exhibit chaos, although most lasers do not operate at the parameter values which lead to chaos.

*Reference*: Grebogi and Yorke (1997).

**measure**

A real-valued function $\mu$ on a $\sigma$-algebra $\mathcal{B}$ of all subsets of $B$ is a measure if:

(i) $\mu(\emptyset) = 0$;

(ii) $\mu(B) \geq 0$ for all $B \in \mathcal{B}$;

(iii) $\mu(\cup_k B_k) = \sum_k \mu(B_k)$ if $\{B_k\}$ is a finite or infinite sequence of pairwise disjoint subsets of $\mathcal{B}$, i.e. is $B_i \cap B_j = \emptyset$ for $i \neq j$.

*Remark*: This definition does not exclude the possibility that $\mu = \infty$ for some $B \in \mathcal{B}$.

Let $\mathcal{B}$ be a $\sigma$-algebra of subsets of $X$, and if $\mu$ is a measure on $\mathcal{B}$, then the triple $(X, \mathcal{B}, \mu)$ is called a measure space.

The sets belonging to $\mathcal{B}$ are called measurable sets.

Let $(X, \mathcal{B}, \mu)$ be a measure space. A real-valued function $f : X \to \mathcal{R}$ is measurable if $f^{-1}(\Delta) \in \mathcal{B}$ for every interval $\Delta \subset \mathcal{R}$.

*References*: Edgar (1990); Lasota and Mackey (1985).

**measure-preserving**
Let $(X_1, \mathcal{B}_1, m_1)$, and $(X_2, \mathcal{B}_2, m_2)$ be probability spaces.

A transformation $T : X_1 \to X_2$ such that

$$T^{-1}(\mathcal{B}_2) \subset \mathcal{B}_1$$

i.e.

$$B_2 \in \mathcal{B}_2 \Rightarrow T^{-1}B_2 \in \mathcal{B}_1$$

is termed measurable.

A transformation $T : X_1 \to X_2$ is measure-preserving if $T$ is measurable and

$$m_1(T^{-1}(B_2)) = m_2(B_2)$$

for all $B_2 \in \mathcal{B}_2$. We say that the transformation $T : X_1 \to X_2$ is an invertible measure-preserving transformation if $T$ is measure-preserving, bijective, and $T^{-1}$ is also measure-preserving.

*Reference*: Arnold and Avez (1968).

**measure-preserving maps** (*see* **volume preserving maps**)

**measure theoretic entropy**
Let $(X, \mathcal{B}, m)$ be a probability space, and let

$$T : (X, \mathcal{B}, m) \to (X, \mathcal{B}, m)$$

be an invertible measure-preserving transformation. If we now let $\mathcal{A}$ be a finite sub-algebra of $\mathcal{B}$ such that

$$\bigcap_{n=-\infty}^{\infty} T^n \mathcal{A} = \mathcal{B}$$

then

$$h(T) = h(T, \mathcal{A})$$

where

$$h(T) = \sup_{\mathcal{A}} h(T, \mathcal{A}).$$

In this case $h(T)$ is called a measure theoretic entropy.

*Reference*: Walters (1982).

**Meissner's equation**
The ordinary differential equation

$$\frac{\mathrm{d}^2x}{\mathrm{d}t^2} + a^2 f(t)x = 0$$

and where

$$f(t) = \begin{cases} 1 & : \quad 0 \leq t < \frac{1}{2} \\ -1 & : \quad \frac{1}{2} \leq t < 1 \end{cases},$$

and $f$ has period 1, for $a > 0$ is called Meissner's equation.

*Reference*: Drazin (1992).

**Melnikov method**
This is an analytical method that can be useful in investigations of chaotic behaviour. The main idea behind this method was proposed by Melnikov in 1963. He considered an 'unperturbed' system of a planar ordinary differential equation

$$\frac{\mathrm{d}x}{\mathrm{d}t} = f(x)$$

where $x \in \mathcal{R}^n$, to have a hyperbolic fixed point connected to itself by a homoclinic orbit. Then the perturbed system with a time-periodic excitation is such that the hyperbolic fixed point becomes a hyperbolic periodic orbit, whose stable and unstable manifolds may intersect transversely. In many systems this intersection is a necessary condition for chaotic dynamics.

*References*: Guckenheimer and Holmes (1983); Melnikov (1963); Wiggins (1988).

**Menger sponge**
Consider the unit cube subdivided into 27 smaller cubes by trisecting the edges. Now remove the centre cube and the 6 cubes in the centre of the faces; this means that 20 cubes remain. The boundary of these 20 cubes must also remain, so that the set will be compact. We continue in the same way by dividing the smaller cubes, etc. The Menger sponge is the limit set of this process. The first steps in the construction of the Menger sponge are shown in Figure 88.

*Reference*: Steeb (1991).

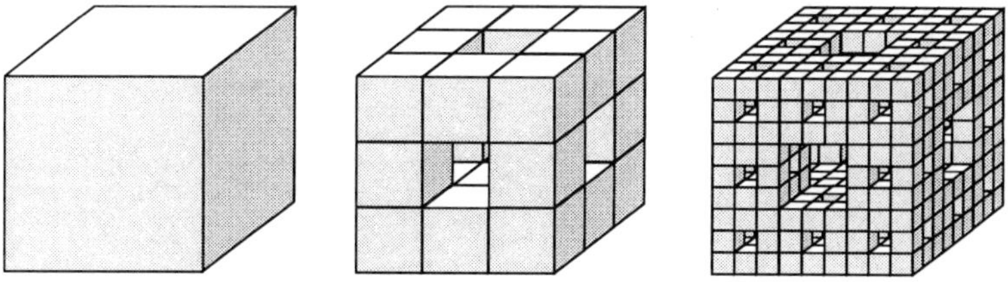

**Figure 88** First steps in the construction of the Menger sponge

**metric space**

A metric space $A$ is a set, together with a distance-function $d : A \times A \to \mathcal{R}$, which satisfies the following conditions:

(i) $\mathrm{d}(x,y) \geq 0$ with equality if and only if $x = y$;

(ii) $\mathrm{d}(x,y) = \mathrm{d}(y,x)$;

(iii) $\mathrm{d}(x,y) + \mathrm{d}(y,z) \geq \mathrm{d}(x,z)$, - namely the triangle inequality.

**mixing**

The endomorphism $T$ is mixing if, for any two measurable sets $A_0$ and $A_1$,

$$\lim_{n\to\infty} \mu(T^{-k_n^0} A_0 T^{-k_n^1} A_1) = \mu(A_0)\mu(A_1)$$

for any non-negative integer sequences $(k_1^0, k_2^0, ...)$, $(k_1, k_2, ...)$, satisfying

$$\lim_{n\to\infty} |k_n^0 - k_n^1| = \infty.$$

**Möbius strip**

The non-orientable manifold $M \subset \mathcal{R}^3$, with a boundary formed by gluing the ends of a strip after inserting a twist (shown in Figure 89), is called a Möbius strip.

**mode locking**

Consider a periodically forced nonlinear system and choose the frequency and amplitudes of forcing as control parameters. When the amplitude of the forcing signal is zero, the frequency of the driven system is independent of the driving frequency. As the amplitude of the forcing is increased, one finds bands of frequencies over which the driving frequency and the system response frequencies are related by rational numbers. This phenomenon is called mode (or frequency) locking.

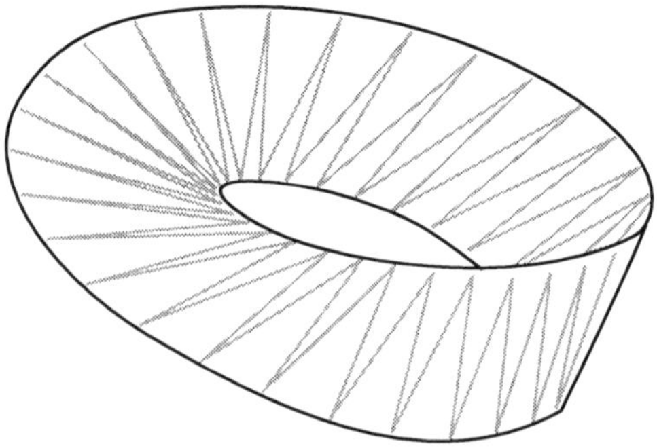

**Figure 89** A Möbius strip

*Reference*: Nayfeh and Mook (1979).

**monodromy matrix**

Let

$$\frac{du}{dt} = A(t)u \tag{1}$$

where $u \in \mathcal{R}^n$ and $A(t)$ is a $(n \times n)$-dimensional matrix of periodic functions with period $T$, be a system of ordinary differential equations. By classical Floquet theory, any fundamental matrix $\Phi$, which is defined as a non-singular matrix satisfying

$$\frac{d\Phi}{dt} = A(t)\Phi(t)$$

can be given as

$$\Phi(t) = P(t)e^{(tR)}.$$

$P(t)$ is a non-singular matrix of periodic functions with the same period $T$, and $R$ is a constant matrix, whose eigenvalues $(\lambda_1, ..., \lambda_n)$ are called characteristic exponents of the dynamical system (1).

Given a choice of the fundamental matrix $\Phi(t)$, we have

$$\exp(TR) = \Phi(t_0)\Phi(t_0 + T)$$

which does not depend on the initial time $t_0$. The matrix $M(T) = \exp(TR)$ is called the monodromy matrix of (1).

**Morse–Smale system**

This system is one for which:

(i) the number of fixed points and periodic orbits is finite and each is hyperbolic;

(ii) all stable and unstable manifolds intersect transversally;

(iii) a non-wandering set consists only of fixed points and periodic orbits.

*Reference*: Guckenheimer and Holmes (1983).

**Moser's theorem**
This theorem shows that in a neighbourhood of a transverse homoclinic point there exists an invariant Cantor set on which the dynamics are topologically conjugate to a full shift on $N$ symbols.

*Reference*: Wiggins (1990).

**Moser's twist theorem**
This theorem guarantees the existence of invariant circles in area-preserving maps.

Consider the unperturbed integrable map

$$I \to I \tag{1}$$

with $\theta \to \theta + \alpha(I)$, defined on the annulus

$$A = \{(I, \theta) \in \mathcal{R}^+ \times \mathcal{S}^1 | I \in [I_1, I_2]\}$$

where $\mathcal{S}^1$ is a one-dimensional circle and the perturbed map

$$I \to I + f(I, \theta) \tag{2}$$

in which

$$\theta \to \theta + \alpha(I) + g(I, \theta)$$

and with $f$ and $g$ also defined on $A$. In order for (2) to be a perturbation of (1), $f$ and $g$ must be small. Let $C^r(A)$ denote the class of $C^r$ functions defined as follows:

$$h \in C^r(A) \to |h|_r = \sup_{A, i+j \le r} \left| \frac{\partial^{i+j} h}{\partial I^i \partial \theta^j} \right| .$$

*Moser's twist theorem*: Let $\epsilon > 0$ be a positive number with $\alpha(I) \in C^r$, $r \geq 5$, and $|\partial\alpha/\partial I| \geq \nu > 0$ in $A$. Then there exists a $\delta$, depending on $\epsilon, r$, and $\alpha(r)$, such that (2) with $f, g \in C^r(A)$, $r \geq 5$, and

$$|f(I, \theta) - I|_r + |g(I, \theta) - \alpha(I)|_r < \nu\delta$$

possesses an invariant circle in $A$ with the parametric representation

$$I = \bar{I} + u(t), \qquad \theta = t + v(t), \qquad t \in [0, 2\pi)$$

where $u$ and $v$ are $C^1$ with period $2\pi$ and satisfy

$$|u|_1 + |u|_2 < \epsilon$$

with $\bar{I} \in [I_1, I_2]$. Moreover, the map restricted to this invariant circle is given by

$$t \to t + \omega, \qquad t \in [0, 2\pi)$$

where $\omega$ is incommensurate with $2\pi$ and satisfies the infinitely many conditions

$$\left| \frac{\omega}{2\pi} - \frac{p}{q} \right| \geq \gamma q^{-\tau} \tag{3}$$

for some $\gamma, \tau > 0$ and all integers $p, q > 0$. In fact, each choice of $\omega \in [\Omega(I_1), \Omega(I_2)]$ satisfying (3), gives rise to such an invariant circle.

*Reference*: Wiggins (1990).

**multifractals**
Multiscale, non-uniform fractals are called multifractals.

*Remark*: Fractals found in nature or complicated dynamical systems are statistically but not geometrically self-similar. A geometrically self-similar fractal is constructed by iterating a single first-generation length $l$ repeatedly, so that all of the $l_n$ have the same length in the $n$th-generation (for example, *see* **Koch curve**). A multifractal is constructed by repeated iteration of two or more first-generation length scales and the result is a non-uniform fractal.

*Reference*: Peitgen *et al.* (1992).

**multiple scales method**
This is a perturbation method for the approximate solution of ordinary differential equations which uses a representation of the response as a function of multiple independent variables, or scales.

*Reference*: Nayfeh (1985).

**multistability** (*see* **bistability**)

# N

**natural measures**

An ergodic invariant measure is one whose initial set has positive Lebesgue measure, i.e. if $\mu$ is a natural measure, and $\phi_t(x_0)$ a flow of the dynamical system, such that $f \in L^1$, then

$$\lim_{T\to\infty} \frac{1}{T} \int_{t=0}^{T} f(\phi_t(x))\mathrm{d}t = \int_x f\mathrm{d}\mu.$$

for a positive measure set of initial conditions $x_0$.

*Remark*: Natural measures have associations with experimentally measurable quantities such as fractal dimension and Lyapunov exponents. They are time averages of dynamical response and generally fall into the area of mathematics that we call ergodic theory. One such measure is the so-called SRB measure, named after Sinai, Ruelle and Bowen.

*Reference*: Lasota and Mackey (1985).

**Navier–Stokes equation**

The partial differential equation which governs the velocity field in the flow of linear, incompressible, viscous Newtonian fluid, given by

$$\frac{\partial u}{\partial t} + (u\nabla)u = -\nabla p + \frac{1}{R_e}\nabla^2 u$$

where $u$ is the velocity vector, $\nabla$ the gradient operator, $\rho$ the density, $p$ the pressure, $\mu$ the viscosity, and $R_e$ the Reynolds number, is called a Navier–Stokes equation.

**negative damping** (*see* **damping**)

**negative-resistance oscillator**

This is the basic oscillator used in electronic circuit shown schematically in Figure 90(a), where the inductor and capacitor are assumed to be linear, time invariant and passive, i.e. $L > 0$ and $C > 0$. The resistive element in an active circuit is characterised by the voltage-controlled function $i = h(V)$, shown in Figure 90(b).

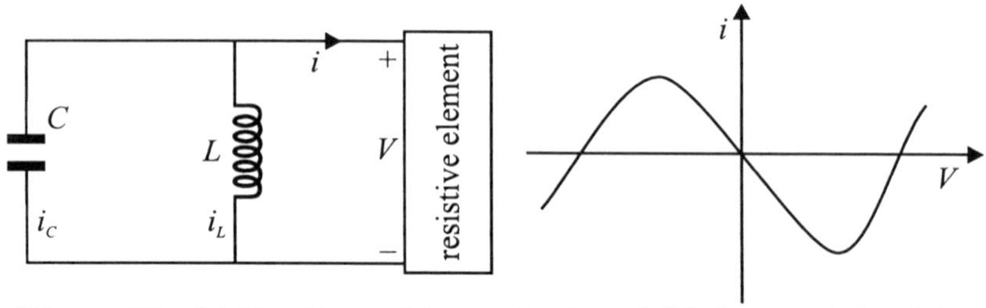


**Figure 90** (a) Negative-resistance circuit and (b) characteristic of the resistive element

By using Kirchoff's current law one obtains

$$i_C + i_L + i = 0$$

and hence

$$C\frac{\mathrm{d}V}{\mathrm{d}t} + \frac{1}{L}\int_{-\infty}^{t} V(s)\mathrm{d}s + h(V) = 0.$$

Differentiating with respect to the time and multiplying by $L$ yields

$$CL\frac{\mathrm{d}^2 v}{\mathrm{d}t^2} + V + L\frac{\mathrm{d}h}{\mathrm{d}V}\frac{\mathrm{d}V}{\mathrm{d}t} = 0$$

If we use the change of variables, $\tau = \frac{t}{\sqrt{CL}}$, and $\epsilon = \sqrt{L/C}$ the above equation produces

$$\frac{\mathrm{d}^2 V}{\mathrm{d}\tau^2} + \epsilon\frac{\mathrm{d}h}{\mathrm{d}V}\frac{\mathrm{d}V}{\mathrm{d}\tau} + V = 0$$

which is a special form of Liénard's equation

$$\frac{\mathrm{d}^2 V}{\mathrm{d}t^2} + f(V)\frac{\mathrm{d}V}{\mathrm{d}t} + g(V) = 0.$$

When $h(V) = -V + V^3/3$ the current takes the form

$$\frac{\mathrm{d}^2 V}{\mathrm{d}t^2} - \epsilon(1 - V^2)\frac{\mathrm{d}V}{\mathrm{d}t} + V = 0$$

which is the well-known van der Pol equation.

**neighbourhood**
A neighbourhood of a point $x$ is a set $U$ which contains $x$ in its interior.

**Neimark bifurcation**
A local bifurcation in which a $T$-periodic limit cycle is replaced by a quasi-periodic trajectory on a $T^2$-torus is called a Neimark bifurcation.

The Neimark bifurcation is effectively a Hopf bifurcation for maps and is sometimes called the Neimark–Sacker bifurcation.

*Reference*: Kuznetsov (1995)

**nerve axon** (*see* **axon**)

**neural networks**
A cellular automata in which each site acting as one neuron can interact with all sites (neurons) in the system is called an artificial neural network or more simply neural network.

Neural networks are used as models for parallel processing.

*Reference*: Rietman (1989).

**Newhouse orbits**
Newhouse showed that for parameter values close to those which result in homoclinic tangencies (*see* **Melnikov method**), infinite sets of stable periodic orbits exist. These orbits are called Newhouse orbits or sinks.

*Reference*: Guckenheimer and Holmes (1983).

**Newton's laws**
There are the three fundamental laws of mechanics developed by Sir Isaac Newton.

1. Every body continues in its state of rest or of uniform rectilinear motion, except if it is compelled by forces acting on it to change that state.

2. The change of motion is proportional to the applied force and takes place in the direction of the straight line along which that force acts.

3. To every action there is always an equal and opposite reaction, or the mutual actions of any two bodies are always equal and oppositely directed along the same straight line.

**Newton method**

Let $f(z)$ be a complex-valued function and $z_0$ be an approximate solution of

$$f(z) = 0.$$

The Newton method is a numerical process to locate the zeros of $f$. The routine finds the next approximation

$$z_{n+1} = z_n - \frac{f(z_n)}{\frac{\mathrm{d}f(z_n)}{\mathrm{d}z_n}}$$

where $n = 0, 1, 2, ...$, etc., provided that $\mathrm{d}f(z_n)/\mathrm{d}z_n \neq 0$.

One calls

$$g(z) = z - \frac{f(z)}{\frac{\mathrm{d}f(z)}{\mathrm{d}z}}$$

the Newton transformation of the function $f(z)$.

*Remark*: The expectation of the Newton method is that a typical orbit $\{f^n(z_0)\}$, which starts from any initial point $z \in \mathcal{C}$, will converge to one of the roots of $f(z)$.

*Reference*: Press *et al.* (1986).

**node** (*see* **fixed points**)

**non-autonomous system**

Consider the following system of differential equations

$$\frac{\mathrm{d}u}{\mathrm{d}t} = f(u, t) \tag{1}$$

where $u(t_0) = u_0, u \in D \subset \mathcal{R}^n$, and $t \in \mathcal{R}^+$. If the right-hand side depends explicitly on time, then (1) is called *non-autonomous*.

*Remark*: An $n$th-order non-autonomous system can always be converted into an $(n+1)$th-order autonomous system by appending an extra variable, e.g.

$$\Theta = \frac{2\pi t}{T}.$$

The autonomous system is given by

$$\frac{\mathrm{d}u}{\mathrm{d}t} = f(u, \Theta), \qquad u(t_0) = u_0$$

and

$$\frac{\mathrm{d}\Theta}{\mathrm{d}t} = \frac{2\pi}{T}, \qquad \Theta(t_0) = \frac{2\pi t_0}{T}.$$

The non autonomous dynamical system

$$\frac{\mathrm{d}u}{\mathrm{d}t} + f(u) = f(t)$$

is called an externally forced system.

The non autonomous dynamical system

$$\frac{\mathrm{d}u}{\mathrm{d}t} + f(u)f(t) = 0$$

is called a parametrically forced system or parametrically excited system.

**non-hyperbolic**
A fixed point $x^*$ of the vector field $f(x) = Ax$, where $x \in \mathcal{R}^n$ and $A$ is an $(n \times n)$-dimensional matrix, is non-hyperbolic if at least one of the eigenvalues of $A$ has a zero real part.

A fixed point of a nonlinear vector field is non-hyperbolic if its linearisation is non-hyperbolic.

**nonlinear function**
A linear function is one which has the same local and global form. Mathematically, a linear function $f(x)$, where $x \in \mathcal{R}$ is one for which $f(x) = ax + b$, where $a$ and $b$ are constants, for all $x$ (strictly $b \neq 0$ is an affine function), whereas a nonlinear function is one which cannot be described in this manner.

Typical examples are the square function $f(x) = x^2$, or the piecewise-linear function, $f(x) = x$ for $x < 1$ and $f(x) = 2x$ for $x \geq 1$.

Nonlinear equations involve terms which are not all linear, e.g.

$$\frac{\mathrm{d}^2x}{\mathrm{d}t^2} + \frac{\mathrm{d}x}{\mathrm{d}t} + x + x^3 = 0.$$

**non-singular map**
A measurable map $S : X \to X$, on a measure space $(X, \mathcal{B}, \mu)$, is non-singular if $\mu(S^{-1}(B)) = 0$ for all $B \in \mathcal{B}$, such that $\mu(B) = 0$.

**non-uniqueness**
The case when the solution of ordinary differential equations is not unique is called non-uniqueness.

Non-uniqueness occurs at bifurcation points.

*Example*: We throw a mass $m$ vertically up in the air along direction $x$. Its vertical velocity is given by

$$m\frac{\mathrm{d}^2x}{\mathrm{d}t^2} = -mg$$

or

$$\frac{m}{2}\left(\frac{\mathrm{d}x}{\mathrm{d}t}\right)^2 = E - mgx$$

where $g$ is the gravitational acceleration and $E$ its potential energy. Therefore the upward velocity of the mass is governed by the equation

$$\frac{\mathrm{d}x}{\mathrm{d}t} = +\sqrt{\frac{2E}{m} - gx}.$$

This equation of motion does not satisfy the Lipschitz condition at $x = 2E/mg$. At this point, this equation will put us on the spurious branch

$$x(t) = \frac{2E}{mg}$$

with $\mathrm{d}x/\mathrm{d}t = 0$, rather than on the physical branch

$$\frac{\mathrm{d}x}{\mathrm{d}t} = -\sqrt{\frac{2E}{m} - gx}.$$

*Reference*: Jackson (1990).

**non-wandering point**
A point $p$ is non-wandering under $f$, if and only if, for any neighbourhood $U$ of $p$ there exists an $n \neq 0$ such that $f^nU \cap U \neq \emptyset$.

The set of all non-wandering points of $f$ is called the non-wandering set $\Omega(f)$.

**normal form**
The normal-form method is based on the use of a nonlinear change of coordinates which simplifies the system as much as possible. To first-order, the normal form is the Jordan normal form. It is particularly used for studying

the properties of a flow (or map) close to bifurcation for which we can reveal the qualitative behaviour of the system.

Given a smooth vector field $f(x)$ on $\mathcal{R}^n$ with $f(0) = 0$, there is a polynomial transformation to new coordinates $y$, such that the differential equation $\mathrm{d}x/\mathrm{d}t = f(x)$ takes the form

$$\frac{\mathrm{d}y}{\mathrm{d}t} = Jy + \sum_{r=2}^{N} w_r(y) + O(|y|^{N+1})$$

where $J$ is the real Jordan form of $A = [\partial f_i/\partial x_j]_{i=j=1}^2|_{x=0}$ and $w_r \in G^r$, $G^r$ is a complementary subspace in $H^r$ of $B^r = L_A(H^r)$, $H^r$ is the real vector space of vector fields whose components are homogeneous polynomials of degree $r$ and $L_A(H^r)$ is an operator which acts on the vector fields in such a way to produce their Lie bracket with $Ay$.

*Reference*: Arrowsmith and Place (1990).

**normal mode**

Consider the linear system

$$\frac{\mathrm{d}x}{\mathrm{d}t} = Ax \tag{1}$$

where $x \in \mathcal{R}^n$. From linear algebra it is known that there exists a real non-singular matrix $T$ such that $T^{-1}AT$ is in the Jordan normal form. If the $n$ eigenvalues are different, then $T^{-1}AT$ is in diagonal form with the eigenvalues as diagonal elements so that the linear transformation $y = Tz$ leads to a simplification. One finds

$$T\frac{\mathrm{d}z}{\mathrm{d}t} = ATz$$

or

$$\frac{\mathrm{d}z}{\mathrm{d}t} = T^{-1}ATz. \tag{2}$$

Usually the Jordan normal form $T^{-1}AT$ is simple and we can integrate (1) immediately, from which $y = Tz$ follows. Components of the vector $z(t)$ are called normal modes.

*Example*: Consider the linear system

$$\frac{\mathrm{d}u_1}{\mathrm{d}t} = -u_1 - 3u_2, \qquad \frac{\mathrm{d}u_2}{\mathrm{d}t} = 2u_2 \tag{3}$$

which can be written in the form (1) with the matrix

$$A = \begin{pmatrix} -1 & -3 \\ 0 & 2 \end{pmatrix}.$$

The eigenvalues of $A$ are $\lambda_1 = -1$ and $\lambda_2 = 2$, with the corresponding eigenvalues being given by

$$\mathbf{v}_1 = \begin{pmatrix} 1 \\ 0 \end{pmatrix}, \qquad \mathbf{v}_2 = \begin{pmatrix} -1 \\ 1 \end{pmatrix}.$$

If we take $T$ to be

$$T = \begin{pmatrix} 1 & -1 \\ 0 & 1 \end{pmatrix}$$

one obtains

$$T^{-1}AT = \begin{pmatrix} -1 & 0 \\ 0 & 2 \end{pmatrix}.$$

Then, under the coordinate transformation $\mathbf{z} = T^{-1}\mathbf{u}$, we obtain

$$\frac{dz_1}{dt} = -z_1, \qquad \frac{dz_2}{dt} = 2z_2$$

which can be solved to obtain normal modes in the form of

$$z_1(t) = c_1 e^{-t}, \qquad z_2(t) = c_2 e^{2t}$$

where $c_1$, $c_2$ depend on the initial conditions. Returning to the original coordinates, $u_1$ and $u_2$, one then obtains

$$u_1(t) = c_1 e^{-t} + c_2(e^{-t} - e^{2t}), \qquad u_2(t) = c_2 e^{2t}.$$

**nowhere dense set**
A subset $X$ of $Y$ is nowhere dense in $Y$ if $X$ has no accumulation points (*see also* **fractal sets**).

# O

**omega-limit set ($\omega$-limit set)**

A point $p$ belongs to an $\omega$-limit set of an orbit $\gamma : \omega(\gamma)$ if $p$ is an $\omega$-limit point (*see* **limit point**).

**on–off intermittency**

Let a set $A$ be an attractor in an $m$-dimensional phase space on a manifold $M$ for all $p$, where $p$ is a control parameter. For $p < p^*$, let $A$ be an attractor on the higher $n$-dimensional phase space, i.e. manifold $N$ for which $M$, $(M \subset N)$ is an invariant manifold. Assume that for $p > p^*$, $A$ is not an attractor in $N$. When $p$ is slightly larger than $p^*$, a typical orbit initiated near the invariant manifold $M$ will eventually drift away from it. However, the orbit can repeatedly revisit the vicinity of the former attractor $A$. This phenomenon is called on–off intermittency.

**open set**

A set $U$ in a metric space is open if for each $x \in U$ there is an $\epsilon > 0$ such that $\mathrm{d}(x, y) < \epsilon$, (where $\mathrm{d}(x, y)$ denote the distance between $x$ and $y$), implies $y \in U$.

**orbit**

Consider a map $f : A \to A$. Then the set

$$\{x, f(x), f^2(x), ...\}$$

where $x \in A$, and $f^n$ denotes the $n$th iteration of map $f$, is called the orbit under $f$.

A trajectory of a flow is also referred to as an orbit.

*Example*: Consider the logistic map $f : [0,1] \to [0,1]$ with $f(x) = 4x(1-x)$ and $x = 1/5$. Then the set $\{1/5, 16/25, 576/625, ...\}$ is the orbit of $1/5$.

*Reference*: Devaney (1989).

**orbital stability** (*see* **stability**)

**ordinary point**
Any point in the phase space of a system

$$\frac{\mathrm{d}x}{\mathrm{d}t} = f(x)$$

where $x \in \mathcal{R}^n$, which is not a fixed point, is said to be an ordinary point.

*Reference*: Arrowsmith and Place (1990).

**organising centre**
This is a term used to describe how qualitatively different types of bifurcation can interact, thus leading to more complex behaviour.

*Reference*: Mullin (1993).

**orientation-preserving**
A homeomorphism $f : X \to Y$ is orientation-preserving if a right-handed coordinate system in $X$ is mapped into a right-handed system in $Y$.

*Example*: Orientation-preserving and non-orientation-preserving homeomorphisms are illustrated in Figure 91.

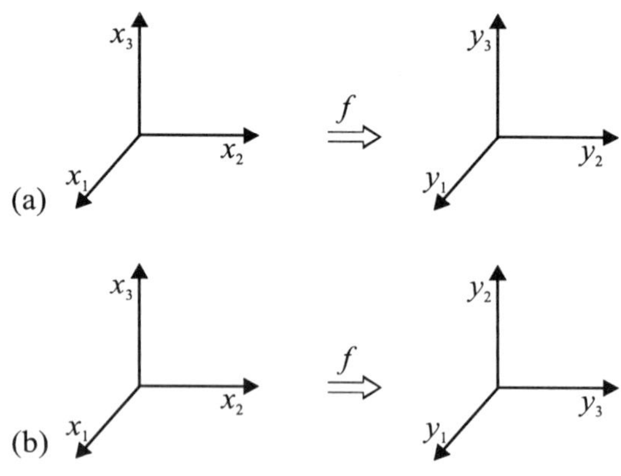


**Figure 91** (a) Orientation-preserving and (b) non-orientation-preserving homeomorphisms

A diffeomorphism $f : \mathcal{R}^3 \to \mathcal{R}^3$ is orientation preserving if the Jacobian is positive, i.e.

$$\det |\frac{\partial f_i}{\partial x_j}| > 0$$

at every point $x \in \mathcal{R}^3$.

*Remark*: If the phase portraits of two dynamical systems

$$\frac{\mathrm{d}x}{\mathrm{d}t} = f(x)$$

with $x \in \mathcal{R}^3$, and

$$\frac{\mathrm{d}y}{\mathrm{d}t} = g(y)$$

with $y \in \mathcal{R}^3$, can be related by an orientation-preserving homeomorphism, then the systems are topologically orbitally equivalent.

**Oseledec theorem**
This theorem states that Lyapunov exponents are well defined for a very wide class of systems.

Let $T$ be a measure-preserving transformation of the probability space $(X, \mathcal{B}, m)$. Let

$$A : X \to L(\mathcal{R}^k, \mathcal{R}^k)$$

be measurable and suppose that

$$\ln ||A(x)|| \in L^1(m)$$

where $L^1(m)$ denotes the space of the functions $f$ with $|f|$ integrable in the sense of Lebesgue. There exists $B \in \mathcal{B}$, with $TB \subset B$ and $m(B) = 1$, with the following properties:

(i) There is a measurable function $s : B \to \mathcal{Z}^+$ with

$$s \cdot T = s.$$

(ii) If $x \in B$ there are real numbers

$$\lambda^{(1)}(x) < \lambda^{(2)}(x) < \ldots < \lambda^{(s(x))}(x)$$

where $\lambda^{(1)}$ could be $-\infty$.

(iii) If $x \in B$ there are linear subspaces of $\mathcal{R}^k$, i.e.

$$\{0\} = V^{(0)} \subset V^{(1)} \subset \ldots \subset V^{(s(x))} = \mathcal{R}^k.$$

(iv) If $x \in B$ and $1 \le i \le s(x)$ then

$$\lim_{n\to\infty} \frac{1}{n} \ln ||A(T^{n-1}x) \cdot \ldots \cdot A(Tx) \cdot A(x)v|| = \lambda^{(i)}(x)$$

for all $v \in V^{(i)}(x) \setminus V^{(i-1)}(x)$.

(v) The functions $\lambda^{(i)}$ are defined and measurable on $\{x : s(x) \ge i\}$ and

$$\lambda^{(i)}(Tx) = \lambda^{(i)}(x)$$

on this set.

(vi) If $s(x) \ge i$, then

$$A(x)(V^{(i)}(x)) \subset V^{(i)}(Tx).$$

*Remark 1*: It should be noted that the most important assumption is (iv), while the others are only technicalities.

*Remark 2*: This theorem is also called a multiplicative ergodic theorem.

The numbers $\lambda^{(1)}(x), \ldots, \lambda^{(s(x))}$ are the Lyapunov exponents.

*References*: Oseledec (1968); Steeb (1991).

**outset**
This is a term used for the unstable manifold of a saddle fixed point.

*Reference*: Thompson and Stewart (1986).

# P

**Painlevé's theorem**
If the only singularities of the general solution of the ordinary differential equation

$$\frac{\mathrm{d}x}{\mathrm{d}t} = f(x, z) \qquad (1)$$

where

$$x \in \mathcal{R}^n \text{and } z \in \mathcal{C}$$

in the complex $z$-plane are movable poles (singularities in a complex plane which depend on the initial conditions) of any finite order, then we say that equation (1) has the Painlevé property and is a $P$-type ordinary differential equation.

Any system of type (1) which has the Painlevé property is integrable.

*Remark*: There exist integrable systems of ordinary differential equations which do not have the Painlevé property.

The Painlevé test which gives the necessary, but not the sufficient conditions for equation (1) to have the Painlevé property consists of three steps:

(i) Find the dominant singular behaviour.

(ii) Find the resonances, i.e. the powers (degrees) associated with the arbitrary coefficients.

(iii) Determine all of the constants of integration.

*Reference*: Steeb and Euler (1988).

**parametric excitation** (*see* **nonautonomous system**)

**parametrically forced system** (*see* **nonautonomous system**)

**partial differential equation (PDE)**
A relationship involving one or more functions of several variables and their partial derivatives is called a partial differential equation, e.g. the heat equation, which typically also has associated boundary conditions.

**Peano curve**
This curve is the limit of the following construction. At the first step we have a line segment. Going from one step to another, each line segment is replaced by nine segments with one third of the length, in the way shown in Figure 92. In the limit, the Peano curve fills the plane.

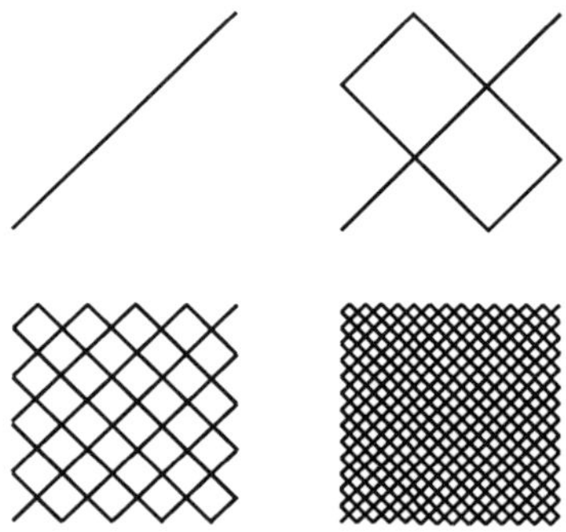

**Figure 92** First steps in the construction of the Peano curve

*Reference*: Becker and Dorfler (1989).

**Peixoto's theorem**
Based on the earlier work of Poincaré, M.M. Peixoto developed a theorem for the structural stability of two-dimensional flows, and in particular guaranteed that they typically contain only invariant sets which are either sources, sinks, saddles or closed orbits (repelling or attracting).

*Theorem*: A $C^r$ vector field on a compact two-dimensional manifold $M^2$ is structurally stable, if and only if the following apply:

(i) the number of fixed points and closed orbits is finite, and each is hyperbolic;

(ii) there are no orbits connecting saddle points;

(iii) the non-wandering set consists of only fixed points and periodic orbits.

Moreover, if $M^2$ is orientable, the set of structurally stable vector fields is open-dense in the set of all $C^r$ vector fields on two-dimensional manifolds.

*Reference*: Guckenheimer and Holmes (1983).

**pendulum**
The motions of a pendulum of mass $m$ and length $l$ (shown in Figure 93) have been used often to study dynamical behaviour.

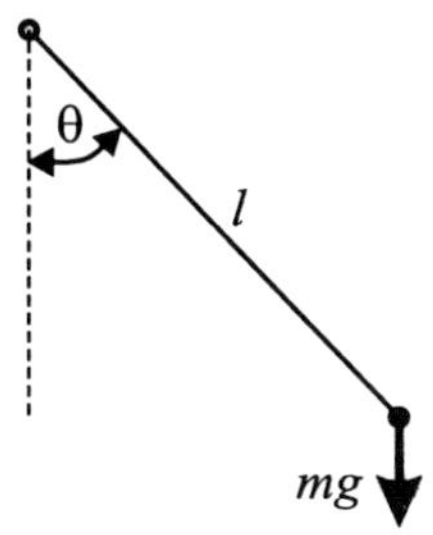


**Figure 93** Pendulum

The motions of the simple pendulum can be described by the equation

$$ml\frac{\mathrm{d}^2\theta}{\mathrm{d}t^2} + mg\sin\theta = 0$$

with a natural frequency $\omega = \sqrt{g/l}$.

**periodic function**
If there exists a $T > 0$ such that a function $f(t)$ fulfills the following condition:

$$f(t+T) = f(t)$$

for all $t$, then the function $f(u,t)$ is said to be time-periodic with period $T$.

**periodic motion**
This type of motion is one for which the solution describing the motion $x(t)$ is such that

$$x(t+T) = x(t)$$

for some time interval $T$, where $T$ is the minimal interval of this property, and is called the period of the motion.

**period-doubling bifurcation**
A bifurcation in which a period-$T$ response of a dynamical system is replaced by a period-$2T$ solution, or more generally a period $- 2^nT$ $(n = 0, 1, ...)$

solution, and which is then further replaced by a period – $2^{n+1}T$ solution, is called a period-doubling bifurcation.

*Example*: Consider the dynamics of the van der Pol equation:

$$\frac{dx_1}{dt} = x_2$$

$$\frac{dx_2}{dt} = -\mu(x_1^2 - 1)x_2 - \omega_0^2 x_1 + p\cos x_3$$

and

$$\frac{dx_3}{dt} = \omega. \tag{1}$$

The bifurcation diagram in Figure 94(a) shows the projections of the attractors in the Poincaré cross-section onto the coordinate $x_1$ for a variation in $\omega$. The plot of the largest non-zero Lyapunov exponent versus $\omega$ is shown in Figure 94(b). Starting with the following values of the system parameters, $\mu = 5$, $\omega_0 = 1$, $p = 5$ and $\omega = 2.457$, we observe a period-four oscillation.

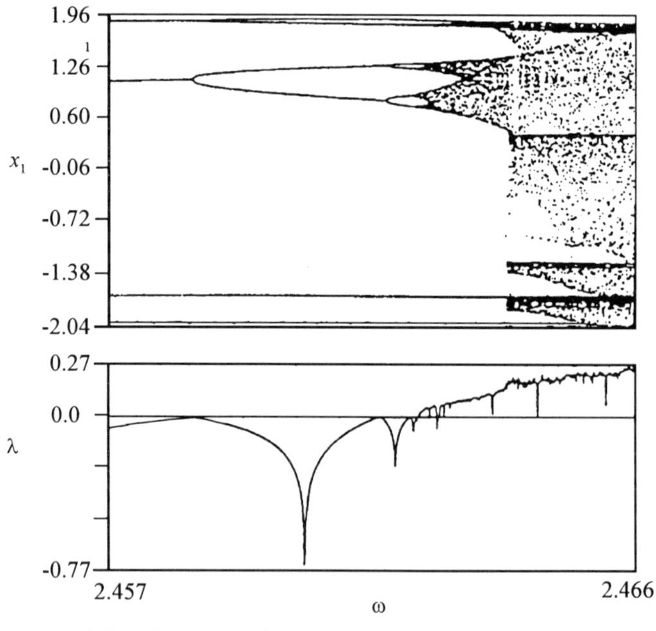


**Figure 94** (a) Bifurcation diagram of the van der Pol's equation, and (b) plot of the largest non-zero Lyapunov exponent versus $\omega$

With a further increase of $\omega$, the system undergoes period-doubling bifurcations. Examples of $4 \times 2^n$, $(n = 0, 1, 2)$, periodic oscillations are shown, respectively, in Figure 95(a–c). This cascade of period-doubling bifurcations culminates in the chaotic attractor shown in Figure 95(d).

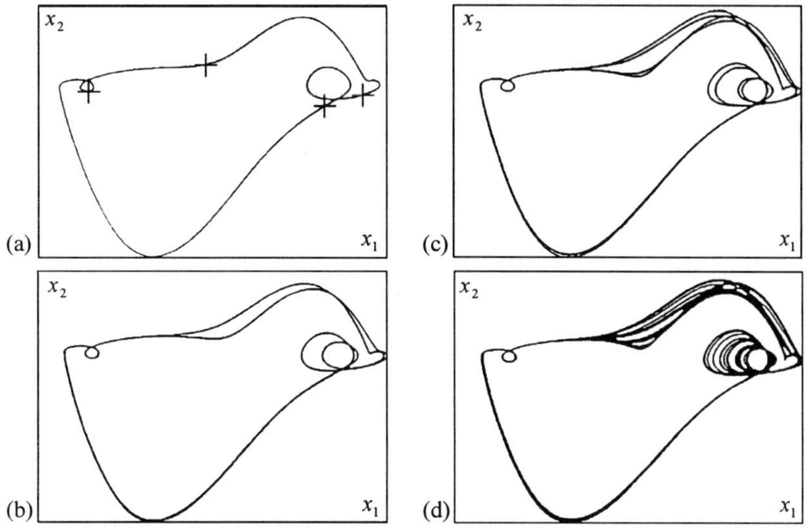


**Figure 95** Period-doubling cascade (van der Pol equation): (a) $\omega = 2.457$; (b) $\omega = 2.460$; (c) $\omega = 2.462$; (d) $\omega = 2.463$

For another example of period-doubling bifurcation, *see* **logistic map**.

*Remark*: The period-doubling bifurcation is sometimes also referred to as a flip-bifurcation.

*References*: Feigenbaum (1978); Thompson and Stewart (1986).

**period-$n$ point**

Consider a map $x_{i+1} = f(x_i)$. A point $x$ is said to be a periodic point of period $n$ if

$$x = f^n(x)$$

and $f^k(x) \neq x$ for $k < n$.

The non-periodic point $x^*$ is said to be eventually periodic if

$$x = f^m(x^*)$$

is a periodic point for some finite $m$.

*Example 1*: For $n = 1$, $x$ is a fixed point.

*Example 2*: For the logistic map (*see* **logistic map**) with $a = 3$, the point

$$x^* = 1/3$$

is eventually periodic with $m = 1$.

*Reference*: Devaney (1989).

**perturbation analysis**
This form of analysis may be used to assess the stability of fixed points and periodic solutions. Consider the system

$$\frac{\mathrm{d}x}{\mathrm{d}t} = f(x)$$

where $x \in \mathcal{R}^n$, which exhibits an equilibrium point at $x = x^*$ such that $f(x^*) = 0$. The stability is examined by superimposing a small perturbation, $\zeta$. Replacing $x$ by $x^* + \zeta$, expanding $f(x)$ as a Taylor series and ignoring terms of $\zeta^2$ and higher, leads to the variational equation

$$\frac{\mathrm{d}\zeta}{\mathrm{d}t} = \zeta \frac{\mathrm{d}f(x)}{\mathrm{d}x}|_{x=x^*} = J\zeta.$$

Therefore, for stability we require the real parts of the eigenvalues of the Jacobian matrix $J$ to be negative (*see also* **linearisation**).

**perturbation methods**
These methods are used for finding approximate solutions to differential equations which usually involve a small parameter. For example consider the equation

$$\ddot{x} + x = \epsilon f(x, \dot{x}) \tag{1}$$

where $\epsilon$ is called the perturbation.

Various methods exist to find approximate solutions $x(t, \epsilon)$ including the Lindstedt-Poincaré method and the method of multiple scales. Typically substitution of a trial solution into the equation (1) yields an expression from which coefficients in powers of $\epsilon$ can be equated with a truncation of higher order terms (since $\epsilon$ is small).

**Pesin equality**
This states that when the measure is smooth, the measure theoretic entropy is then equal to the sum of the positive Lyapunov exponents (*see also* **variational principle for entropy**).

**phase portrait** (*see* **phase space**)

**phase space**
Consider the equation

$$\frac{dx}{dt} = f(x) \tag{1}$$

where $x \in \mathcal{D} \subset \mathcal{R}^n$. The set $\mathcal{D}$ is called the phase space.

Generally, phase space is the collection of variables needed to have a deterministic evolution.

The solutions of (1) for different initial conditions generate a family of oriented phase curves in the phase space, with the former being called a phase portrait.

The phase portrait of a solution is a plot in phase space of the orbit evolution.

A global phase portrait illustrates both steady states and transients leading to any steady state.

**piecewise-linear function**
Consider a function $f(x)$, where $x \in (a, b) \subseteq \mathcal{R}$. If the interval $(a, b)$ can be divided into sub-intervals $(a, c_1), (c_1, c_2), ..., (c_n, b)$ $(n \geq 1)$, in such a way that in each sub-interval the function $f(x)$ is affine, then $f(x)$ is called piecewise-linear.

**pitchfork bifurcation** (*see* **bifurcation**)

**Plykin attractor**
This is the simplest known non-trivial planar hyperbolic attractor, being named after R. Plykin who published its details in 1974. A modified version consists of a compact subset $D \subset \mathcal{R}^2$ with three holes which each contain a source, as shown in Figure 96.

In each hole we define $f : D \to \mathcal{R}^2$ geometrically so that $f(D)$ lies inside the interior of $D$ as shown. The attractor is $A = \cap_{n \leq 0} f^n(D)$.

*References*: Guckenheimer and Holmes (1983); Newhouse (1980); Plykin (1974).

**Poincaré–Andronov–Hopf bifurcation** (*see* **Hopf bifurcation**)

**Poincaré–Bendixon theorem**
Suppose that the trajectory $\Phi_t(u_0)$ of the differential equation $du/dt = f(u)$, where $u \in \mathcal{R}^2$, with flow $\Phi_t$, is contained in a bounded region $D$ of the phase space for $t \geq 0$. Then the only possible $\omega$-limit or $\alpha$-limit sets for $\Phi_t(u_0)$ are

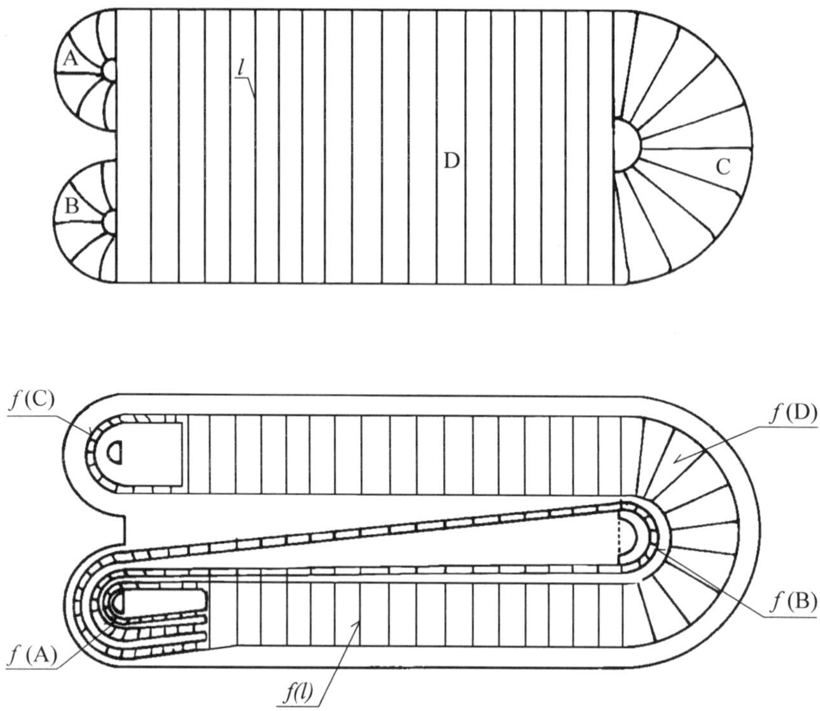


**Figure 96** The Plykin attractor

a critical point or periodic solution.

*Remark 1*: This is only a two-dimensional result.

*Remark 2*: This is a powerful result since it proves the existence of a periodic orbit in $D$ when there are no fixed points in $D$.

There are a number of criteria that guarantee the existence of limit cycles for certain classes of equations. Consider the second-order ordinary differential equation

$$\frac{\mathrm{d}^2u}{\mathrm{d}t^2} + f\left(u, \frac{\mathrm{d}u}{\mathrm{d}t}\right)\frac{\mathrm{d}u}{\mathrm{d}t} + g(u) = 0. \tag{1}$$

Provided that the following conditions hold, then equation (1) has at least one limit cycle.

(i) $ug(u) > 0$ for all $u > 0$;

(ii) $\int_0^\infty g(u)\mathrm{d}u = \infty$;

(iii) $f(0,0) < 0$ and there exists a $u_0 > 0$ such that $f(u, \mathrm{d}u/\mathrm{d}t) \geq 0$ for $|u| > u_0$ and every $\mathrm{d}u/\mathrm{d}t$;

(iv) there exists a constant $M > 0$, such that $f(u, \mathrm{d}u/\mathrm{d}t) \geq -M$ for $|u| < u_0$;

(v) there exists a $u_1 > u_0$ such that $\int_{u_0}^{u_1} f(u, \phi(u))\mathrm{d}u \geq Mu_0$

where $\phi$ is an arbitrary positive and monotonically decreasing function of $u$.

*Reference*: Andronov *et al.* (1966).

**Poincaré–Birkoff theorem**

Consider an annulus $A = \{(\theta, \tau)|a \leq \tau \leq b, 0 \leq \theta \leq 2\pi\}$ and an area-preserving twist map $T$, as follows:

$$T : (\theta, \tau) \to (\theta + \alpha(\tau), \tau).$$

We now define the following perturbation to $T$:

$$M_\epsilon(\theta, \tau) \to (\theta + \alpha(\tau) + f(\theta, \tau, \epsilon), \tau + g(\theta, \tau, \epsilon),$$

where $\mathrm{d}\alpha(\tau)/\mathrm{d}\tau \neq 0$, and $f, g$ are $2\pi$-periodic in $\theta$ with

$$f(\theta, \tau, 0) = g(\theta, \tau, 0) = 0.$$

We will suppose that $M_\epsilon$ is defined on an annulus $a \leq \tau \leq b$, $a < b$ and that it is area-preserving for all values of $\epsilon$, i.e.

$$\int_\Gamma \tau \mathrm{d}\theta = \int_{M_\epsilon \Gamma} \tau \mathrm{d}\theta$$

for any closed curve $\Gamma$ in the annulus.

*Theorem*: Given any rational number, $p/q$, between $\alpha(a)/2\pi$ and $\alpha(b)/2\pi$, then there are $2q$ fixed points of $M_\epsilon^q : (\theta, \tau) \to (\theta_q, \tau_q)$ satisfying

$$(\theta_q, \tau_q) = (\theta + 2\pi p, \tau)$$

provided that $\epsilon$ is sufficiently small.

The periodic points which are predicted by this theorem are called Poincaré–Birkhoff periodic points.

*Reference*: Arrowsmith and Place (1990).

**Poincaré index**

Consider a closed curve $\gamma \subset \mathcal{R}^2$ and a vector field $(P, Q)$:

$$\frac{dx}{dt} = P(x, y)$$

and

$$\frac{dy}{dt} = Q(x, y). \tag{1}$$

At each point of $\gamma$, consider the direction of the vector field, as illustrated in Figure 97.

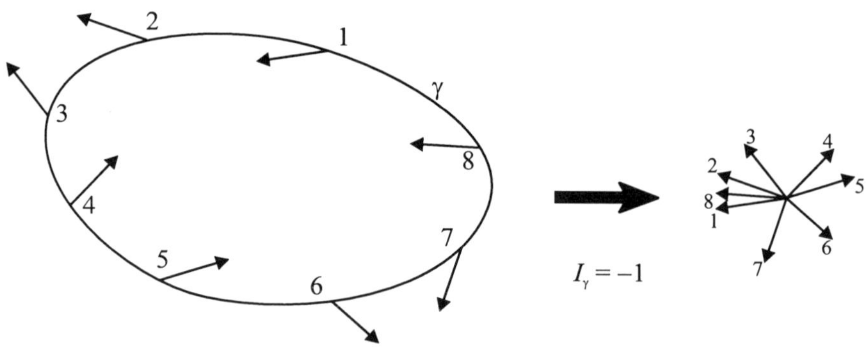


**Figure 97** An example a of Poincaré index

After going around $\gamma$, either this record of field vectors rotates around the intersection point with $\gamma$, or else it does not. The number of rotations of the field vectors about the intersection points with $\gamma$ is called the Poincaré index $I_\gamma$ of the curve $\gamma$, with $I_\gamma > 0$ ($I_\gamma < 0$) if the rotation is in the same (opposite) sense as the rotation carried out along $\gamma$ in the plane.

The Poincaré index, for the vector field determined by (1), can be calculated as

$$I_\gamma = \frac{1}{2\pi} \oint_\gamma d\left(\tan^{-1}\frac{Q}{P}\right) =$$

$$\frac{1}{2\pi} \oint_\gamma \left[\left(P\frac{\partial Q}{\partial x} - Q\frac{\partial P}{\partial x}\right) dx + \left(P\frac{\partial Q}{\partial y} - Q\frac{\partial P}{\partial y}\right) dy\right] [P^2 + Q^2]^{-1}.$$

If $\gamma$ goes around either a single centre, focus or node, then the Poincaré index $I_\gamma = 1$. If it goes around a saddle point, $I_\gamma = -1$.

If $I_\gamma \neq 0$, then there is at least one fixed point in the region enclosed by $\gamma$.

Consider the closed curve $\gamma$ which encloses only one fixed point; the index $I_\gamma$ is then called a Poincaré index of the fixed point.

*Reference*: Jackson (1990).

**Poincaré map**
This map is a classical device due to Henri Poincaré and is used for analyzing dynamical systems. The main idea is to replace the flow of an $n$th-order continuous-time system with an $(n-1)$th order discrete-time system. The latter is constructed by viewing the phase-space diagram stroboscopically on a plane of a section in such a way that the motion is observed not continuously but at a given discrete sequence of times.

The definition of the Poincaré map is slightly different for autonomous and non-autonomous systems. Consider an $n$th-order autonomous system $\mathrm{d}x/\mathrm{d}t = f(x)$, where $x \in \mathcal{R}^n$, and assume that it has a limit cycle $\Gamma$, as illustrated in Figure 98.

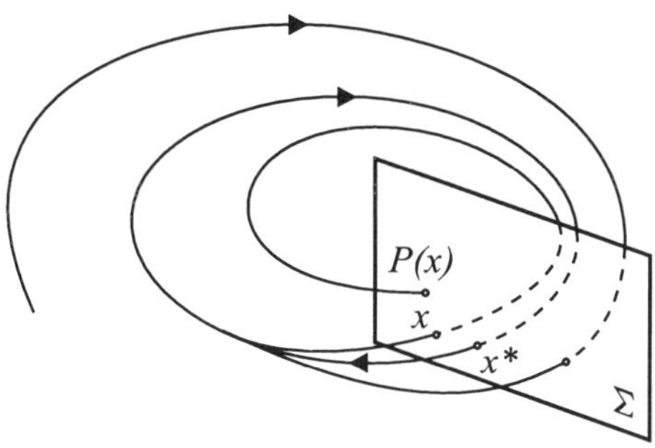


**Figure 98** The Poincaré map for a three-dimensional autonomous system

Strictly speaking, the definition of a Poincaré map does not require the existence of a periodic orbit; however, for illustration let $x^*$ be a point on the limit cycle. Let $\Sigma$ be an $(n-1)$-dimensional surface transverse to $\Gamma$ at $x^*$.

The orbit starting from $x^*$ will cross $\Sigma$ at $x^*$ after a period $T$ of a limit cycle. Trajectories starting on $\Sigma$ in a sufficiently small neighbourhood of $x^*$ will intersect $\Sigma$ in the vicinity of $x^*$. Hence the equation $\mathrm{d}x/\mathrm{d}t = f(x)$ and $\Sigma$ define a mapping $P$ of some neighbourhood $U \subset \Sigma$ of $x^*$ onto another neighbourhood $V \subset \Sigma$ of $x^*$. $P$ is a Poincaré map of the autonomous system.

The transverse surface $\Sigma$ is called the surface of section.

This definition of the Poincaré map is rarely used in simulations and experimental settings because it requires advanced knowledge of the position of a limit cycle. In practice, one chooses an $(n-1)$-dimensional surface $\Sigma$, which divides $\mathcal{R}^n$ into two regions. If $\Sigma$ is chosen properly, then the trajectory under observation will repeatedly pass through $\Sigma$, as shown in Figure 99. A Poincaré map is built out of these crossing points.

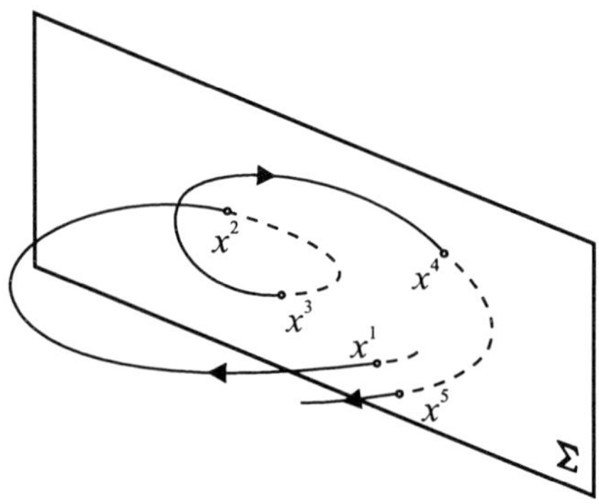


**Figure 99** Practical construction of a Poincaré map for an autonomous system

*Example 1*: Consider a three-dimensional autonomous system. A Poincaré map can be defined as the set

$$\sum = \{(u_1(t), u_2(t)) \quad : \quad t = t_k,\ u_3(t_k) = \text{const}\}.$$

Unfortunately, there is no guarantee that such a map is well-defined, since $\mathbf{u}(t)$ may never intersect $\Sigma$. In the case of a system in Euclidean phase space, with bounded behaviour which does not approach an equilibrium point, there is always some choice of $\Sigma$ for which the Poincaré map is well-defined locally.

Consider now a non-autonomous system of the $n$-th order. For a time-periodic non-autonomous system with a forcing period $T$, which can be transformed into an $(n+1)$th-order autonomous system in the cylindrical phase space $\mathcal{R}^n \times S^1$, the Poincaré map can be defined in the following way. Consider the

$n$-dimensional surface $\Sigma \in \mathcal{R}^n \times S^1$:

$$\Sigma := \left\{ (\mathbf{u}, \theta) \in \mathcal{R}^n \times S^1 : \theta = \theta_0 \right\}.$$

After every period $T$, the orbit $\mathbf{u}(t)$ intersects $\Sigma$ (see Figure 100). The resulting map

$$P : \Sigma \to \Sigma \qquad (\mathcal{R}^n \to \mathcal{R}^n)$$

which maps $\mathbf{u}(t) \to \mathbf{u}(t+T)$ is a globally defined Poincaré map.

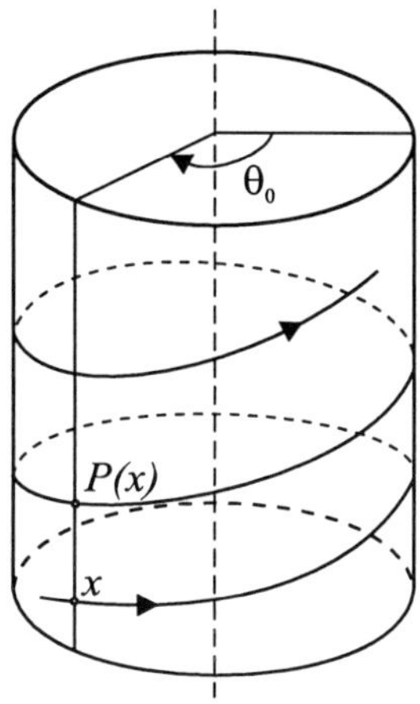


**Figure 100** The Poincaré map of a one-dimensional non-autonomous system

*Example 2*: Consider the forced Duffing equation

$$\frac{\mathrm{d}^2 u}{\mathrm{d}t^2} + a\frac{\mathrm{d}u}{\mathrm{d}t} + bu + cu^3 = F\cos(\Omega t)$$

where $a$, $b$, $c$, $F$ and $\Omega$ are constants. This equation is invariant under the following transformation

$$S : \left\{ \begin{array}{lcl} (u, t) & \to & (u, t + 2\pi/\Omega) \\ (\mathrm{d}u/\mathrm{d}t, t) & \to & (\mathrm{d}u/\mathrm{d}t, t + 2\pi/\Omega) \end{array} \right.$$

and a Poincaré map can be defined as the set

$$\{(u(t), \mathrm{d}u/\mathrm{d}t) \,:\, t = t_0 + 2k\pi/\Omega,\ k = 1, 2, \ldots\}.$$

This map is also sometimes called the return map.

*Reference*: Guckenheimer and Holmes (1983).

**Poisson bracket**
Let $M$ be a differentiable manifold. The Poisson bracket of the two functions $f, g \in C^\infty(M)$ is the function $[f,g] \in C^\infty(M)$, defined by

$$[f,g] = \sum \left( \frac{\partial f}{\partial p_j}\frac{\partial g}{\partial q_j} - \frac{\partial f}{\partial q_j}\frac{\partial g}{\partial p_j} \right).$$

The Poisson bracket is anti-symmetric in $f$ and $g$, i.e. $[f,g] = -[g,f]$.

The Poisson bracket satisfies the Jacobi identity

$$[f,[h,g]] + [h,[f,g]] + [g,[h,f]] = 0$$

for $f, g, h \in C^\infty(M)$.

*Reference*: Arnold (1983).

**polynomial nonlinearities**
Nonlinear functions in the form of a polynomial,

$$P(x) = a_0 + a_1 x + a_2 x^2 + \ldots + a_n x^n,$$

where $n > 1$, describe polynomial nonlinearities.

**population dynamics**
This is the term used to consider the growth in population size of various species which can be modelled either continuously or discretely (*see* **logistic map**, **Ricker map**).

**Poston's catastrophe machine**
The Poston machine is a simple system which may be used to illustrate catastrophe theory. It consists of a uniform wheel of mass $m$ and radius $r$, with a mass $M$ attached at a distance $c \leq r$ from its axis, as shown in Figure 101.

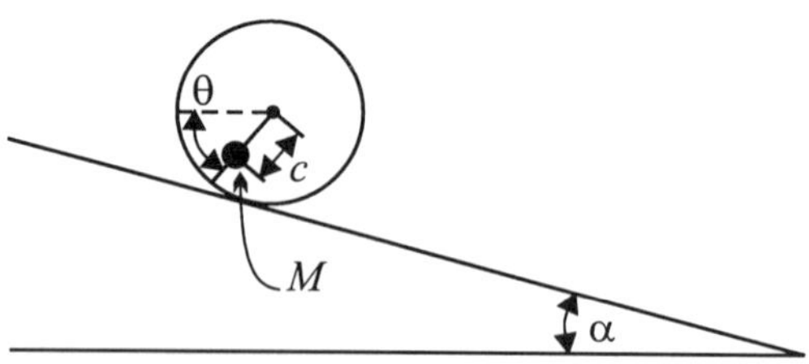


**Figure 101** Poston machine

By changing the parameter $c$, one can observe various catastrophes (*see also* **Zeeman's catastrophe machine**).

*Reference*: Poston and Stewart (1978).

**potential well**
The dynamics of many physical systems can be viewed in terms of the behaviour of a particle moving in a potential well. This is especially true for mechanical systems governed by equations of the form

$$\frac{\mathrm{d}^2x}{\mathrm{d}t^2} + \frac{\mathrm{d}x}{\mathrm{d}t} + f(x) = F(t)$$

where the associated potential-energy function $V(x)$ is found from

$$V(x) = \int f(x)\mathrm{d}x$$

and where $F(t)$ is a small excitation.

**power spectra**
When a function $f(t)$ is not periodic or is quasi-periodic, then it can be expressed in terms of oscillations with a continuum of frequencies. Such a representation is called the Fourier transform of $f$. In this case, the spacing between the frequency components becomes infinitesimally small and the discrete spectrum of the frequency components becomes continuous.

Consider the following transformations in a Fourier series (*see* **spectral analysis**):

$$T \to \infty, \qquad n\omega_0 \to \omega$$

where $\omega$ is a continuous variable, and

$$a_n \to a(\omega)d\omega.$$

This transformation allows the transition from the Fourier series to the Fourier transform.

The appropriate limits lead to the Fourier transform $a(\omega)$ of the function $f(t)$:

$$a(\omega) = \frac{1}{2\pi}\int_{-\infty}^{\infty} f(t)e^{i\omega t}\mathrm{d}t. \tag{1}$$

The inverse transform is given by

$$f(t) = \int_{-\infty}^{\infty} a(\omega)e^{-i\omega t}\mathrm{d}\omega.$$

The Fourier transform $a(\omega)$ can often be complex. Therefore it is useful to define a real-valued function:

$$S(\omega) = |a(\omega)|^2$$

which is called the *power spectrum*. This is a quantity which is very useful in a number of practical applications, for example it allows the main frequencies of the considered system to be determined so that their resonances can be avoided in experiments.

*Example*: Let us consider the solution of the damped linear oscillator

$$\frac{\mathrm{d}^2 u}{\mathrm{d}t^2} + c\frac{\mathrm{d}u}{\mathrm{d}t} + \omega_0^2 u = 0$$

which is given by

$$u(t) = Ae^{-ct}e^{i\omega_0 t}$$

where $c$ is the damping coefficient, $\omega_0$ is the frequency of undamped oscillations and $A$ is a constant which depends on the initial conditions. Application of (1) allows the calculation of the integral for $a(\omega)$:

$$a(\omega) = \frac{A}{2\pi[c - i(\omega - \omega_0)]}$$

and the power spectrum is as follows:

$$S(\omega) = \frac{1}{4\pi^2[c^2 + (\omega - \omega_0)^2]} \; .$$

This power spectrum is given in Figure 102, showing that $S(\omega)$ is symmetric about the dominant frequency $\omega_0$, which is the natural frequency of the undamped system.

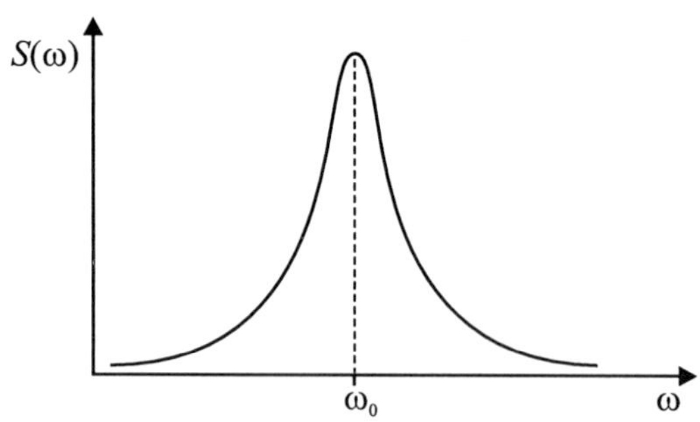


**Figure 102** Power spectrum of the response of a damped linear oscillator

*Reference*: Kaplan (1973).

**predator–prey equations**
These are a set of equations which models the evolutions of species of predators and their prey (*see* **Lotka–Volterra equations**).

**predictability**
This is the ability to determine the future state of a dynamical system, say for $t > t_1$, based on the state for $t < t_1$.

*Remark*: In chaotic systems, lack of practical predictability results from the inability to make physical measurements of initial conditions or numerical computations, with infinite precision. Therefore, the sensitivity due to initial conditions produces the divergence of initially close-by conditions.

**primary resonance**
In continuum systems or systems with many degrees of freedom, such as a vibrating beam, different modes of vibration will exist which can be excited into resonance by the appropriate choice of the frequency of excitation. The primary resonance occurs when the system is excited with frequencies close to its fundamental mode, thus producing the largest amplitudes of response.

**probability density function**
Let the mean value of a random variable $R(\omega)$, in which the basic events $\omega$ are countably specifiable, be given by

$$< R >= \sum_{\omega} P(\omega)R(\omega)$$

where $P(\omega)$ means the probability of the set containing only the single event $\omega$.

In the case of a continuous variable, the probability density function $p(\omega)$ can be defined in the following way. Let $A(\omega_0, d\omega_0)$ be a set

$$(\omega_0 \leq \omega \leq \omega_0 + \mathrm{d}\omega_0)$$

which gives

$$p(\omega_0)\mathrm{d}\omega_0 = P[A(\omega_0, \mathrm{d}\omega_0)] = p(\omega_0, \mathrm{d}\omega_0).$$

*Reference*: Gardiner (1990).

**probability measure**
A measure $\mu$ on a set $A$ is a probability measure if $\mu(A) = 1$ .

**probability space**

Let $X$ be a set. A $\sigma$-algebra of subsets of $X$ is a collection $\mathcal{B}$ of subsets $B$ of $X$ satisfying:

$$X \in \mathcal{B}$$

$$B \in \mathcal{B} \Rightarrow X \setminus \mathcal{B} \in \mathcal{B}$$

and

$$B_n \in \mathcal{B} \Rightarrow \cup_{n=1}^{\infty} B_n \in \mathcal{B}$$

where $n \geq 1$. We call $(X, \mathcal{B})$ a measurable space.

A measure space is a triple $(X, \mathcal{B}, m)$ where $X$ is a set, $\mathcal{B}$ is a $\sigma$-algebra of subsets of $X$, and $m$ is a function $m : \mathcal{B} \to \mathcal{R}^+$ satisfying

$$m\left(\cup_{n=1}^{\infty}\right) = \sum_{n=1}^{\infty} m(B_n)$$

if $\{B_n\}$ is a pairwise-disjoint sequence of elements of $\mathcal{B}$. We say that $(X, \mathcal{B}, m)$ is a probability space if $m(X) = 1$.

*Reference*: Steeb (1991).

**Proudman–Johnson equation**

The partial differential equation

$$\frac{\partial^3 f}{\partial t \partial y^2} = \frac{1}{R}\frac{\partial^4 f}{\partial y^4} + f\frac{\partial^3 f}{\partial y^3} - \frac{\partial f}{\partial y}\frac{\partial^2 f}{\partial y^2}$$

where $f = \mp 1$, and $\partial f / \partial y = 1$ at $y = \pm 1$, is called the Proudman–Johnson equation.

This equation describes the unsteady, non-parallel flow of a viscous fluid in a channel.

*Reference*: Drazin (1992).

**pseudo-orbit** (*see* **$\alpha$-pseudo orbit**)

# Q

**quadratic map**

A one dimensional uni-modal map, written as one of the following

$$x_{n+1} = ax_n - bx_n^2$$

$$x_{n+1} = 1 - ax_n^2$$

or

$$x_{n+1} = c - x_n^2$$

is called the quadratic map.

A further variation of the quadratic map is the logistic map.

**quantum chaology**

Following M. Berry, quantum chaology is the study of semiclassical, but nonclassical, phenomena characteristic of systems whose classical counterparts exhibit chaotic behaviour. 'Semiclassical' means as the Planck constant $h$ tends to zero. The limit is non-trivial because quantum mechanics, considered as depending on the parameter $h$, is essentially singular at the 'classical' origin $h = 0$, in ways that differ from system to system. This non-analytic property is present in all waves in the limits of vanishing wavelengths. The most understood aspects are associated with caustics in integrable systems, and can be expressed in terms of scaling laws involving exponents whose determination involves catastrophe theory. Because of the essential singularity at $h = 0$, the classical limit of quantum mechanics (and also the geometrical-optics limit of electromagnetism) is complicated and conceals a rich variety of phenomena. Quantum theory is a non-perturbative extension of classical mechanics. 'Nonclassical' is incorporated into the definition in order to exclude the trivial sense in which classical unpredictability could be regarded as quantum chaos on the grounds that every classical system is really the $h = 0$ limit of a quantum system.

*References*: Berry (1989); Steeb (1991).

**quantum chaos**
Consider a chaotic Hamiltonian system with a Hamiltonian function $H$. Let $\bar{H}$ be the associated Hamiltonian operator (spectrum). The study of $\bar{H}$ is summarised as quantum chaos.

There is no unique definition of quantum chaos (*see also* **quantum chaology**).

*Reference*: Steeb (1991).

**quasi-linear function**
A function $f(x)$ is called quasi-linear if

$$f(x) = g(x) + \epsilon h(x)$$

where $g(x)$ is linear, $h(x)$ is nonlinear and $\epsilon$ is small.

**quasi-periodic function**
An almost periodic function

$$f(t) = g(\omega_1 t, \omega_2 t, ..., \omega_m t)$$

where $g(.)$ is $2\pi$-periodic in each argument and the frequencies $\omega_k$ are not rationally related, is called a quasi-periodic function.

**quasi-periodic route to chaos** (*see* **route to chaos**)

**quasi-stationary process**
This is a term typically used to describe an experiment in which, for instance, the amplitude of excitation force is held fixed while the frequency is slowly varied, i.e.

$$z(t) = z \cos \omega t$$

where $\omega = \omega_0 + rt$, and $r \ll 1$.

# R

**random dynamical system**

This is a general framework for investigating random systems by analyzing 'non-autonomous' systems where parameters evolve stochastically, which has been particularly studied by L. Arnold and co-workers.

*References*: Arnold (1974); Kloeden and Platen (1992).

**randomly transitional phenomena**

This is a term used by Y. Ueda to describe chaotic motion essentially before the term chaos was universally accepted. The term usefully describes the motion where a system moves close to, and then away from an infinite number of unstable solutions in an apparently random manner. Employing similar concepts, Lorenz has described chaotic motion as the motion of a ball on an infinitely long bagatelle board.

*Reference*: Lorenz (1993); Ueda (1979).

**Rayleigh–Bénard convection**

This occurs when a horizontal layer of fluid is heated from below with a sufficiently large heat flux $Q$ to generate flow. In a classical example, the convection of a fluid between two plates which are held at different temperatures $T$ and $T + \Delta T$ (as shown in Figure 103) is considered. As the temperature difference $\Delta T$ is increased, convective vertical rolls of fluid occur.

*Reference*: Busse (1980).

**Rayleigh equation**

The ordinary differential equation

$$m\frac{\mathrm{d}^2 x}{\mathrm{d}t^2} + \left(-a + b\left(\frac{\mathrm{d}x}{\mathrm{d}t}\right)^2\right)\frac{\mathrm{d}x}{\mathrm{d}t} + kx = 0$$

where $a, b > 0$, is called the Rayleigh equation.

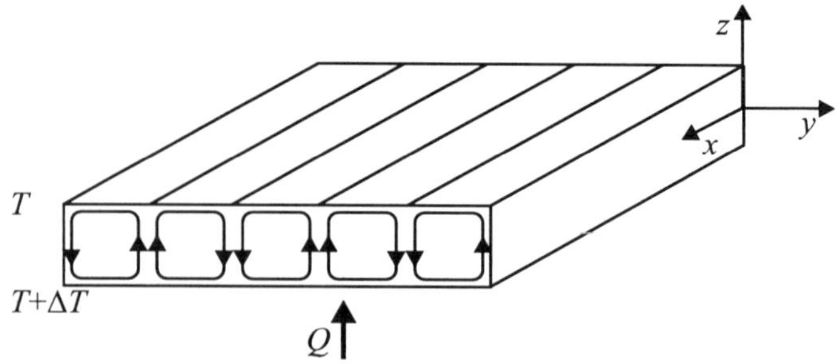


**Figure 103** Illustration of Rayleigh–Bénard convection

**reconstructed phase space**
The phase space constructed from a time series of measurements by using delay coordinates is called the reconstructed phase space (*see also* **embedding**).

**recurrence theorems**
Let $T : X \to X$ be a measure-preserving transformation of a probability space $(X, \mathcal{B}, m)$, and let $E \in \mathcal{B}$ with $m(E) > 0$. Then almost all points of $E$ return infinitely often to $E$ under positive iteration by $T$.

*Remark*: There exists $F \subset E$ with $m(F) = m(E)$, such that for each $x \in F$ there is a sequence, $n_1 < n_2 < n_3 < \ldots < n_i$, of natural numbers with $T^{n_i}(x) \in F$ for each $i$.

Let $g$ be a volume-preserving continuous one-to-one map which maps a bounded region $D$ of Euclidean space onto itself; i.e. $gD = D$. Then on any neighbourhood $U$ of any point of $D$ there is a point $x \in U$ which returns to $U$, i.e. $g^n x \in U$ for some $n > 0$.

*Reference*: Arnold (1988).

**reflection map**
A map $(x, y) \to (x, -y)$, where $x, y \in \mathcal{R}$ produces a reflection about the $x$-axis, is called the reflection map.

**relaxation oscillations**
Consider the van der Pol equation

$$\frac{d^2x}{dt^2} - k(1 - x^2)\frac{dx}{dt} + x = 0.$$

For large $k$, the solution $x(t)$ has the form shown in Figure 104.

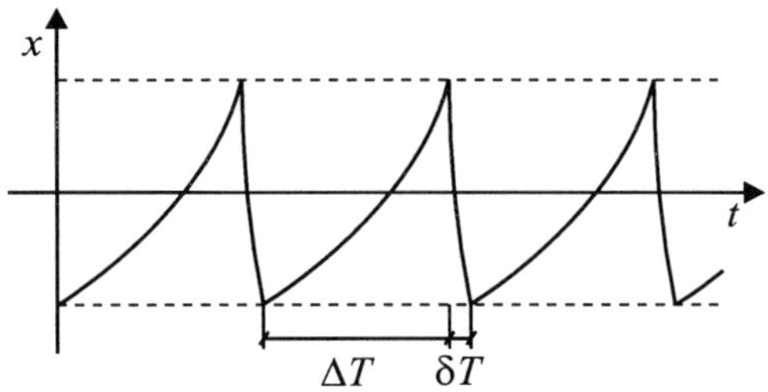


**Figure 104** Relaxation oscillations of the van der Pol equation

The time intervals indicated on the orbit, exhibit regions of fast and slow motion in the phase space. Systems which are self-excited (there is no periodic force), producing periodic response which have such fast (during $\delta T$ intervals) and slowly varying states (during $\Delta T$ intervals) in their periods, are referred to as relaxation oscillations.

*Reference*: Huntley and Johnson (1983).

**renormalisation group theory**
This theory describes semi-groups of scaling transformations, which through theorems such as those of Sharkovskii and Li and Yorke, determine the universal behaviour of a wide class of maps. Feigenbaum used renormalisation group theory to account for the universal scaling of maps during period-doubling.

*Reference*: Guckenheimer and Holmes (1983).

**Renyi dimension**
Let $\mu$ be a probability measure on a $d$-dimensional phase space $X$, where an appropriate partition is chosen that divides the phase space into small cells of volume $D^q$. The Renyi dimensions are defined as

$$\mathcal{D}(q) = \lim_{D^q \to 0} \frac{1}{q-1} \frac{\ln \sum_i p_i^q}{\ln D^q}$$

for $q \in \mathcal{R}$ and $q \neq 1$, and

$$\mathcal{D}(q) = \lim_{D^q \to 0} \frac{\sum_i p_i \ln p_i}{\ln D^q}$$

where, for $q = 1$:

$$p_i = \int_{i\text{th cell}} d\mu(x).$$

*References* Steeb (1991).

**repellor**
An attractor which is unstable is called a repellor (*see* **attractor**).

**resonance**
Consider the forced oscillator

$$\frac{d^2x}{dt^2} + f(x, \frac{dx}{dt}) = A\cos\Omega t.$$

The phenomenon in which the solution $x(t)$ attains a large amplitude at specific values of the driving frequency $\Omega$, is called a resonance.

The principal resonance occurs in the neighbourhood of $\Omega = \omega_n$, and is known as the natural frequency of the oscillations.

The resonance which occurs at a sub-harmonic frequency is called a sub-harmonic resonance.

**resonance curve**
This is used in engineering to visualise the response of a system upon variation of a parameter, and is usually seen as the modulus of the maximum amplitude of displacement $|x|$ versus the variation of the excitation frequency $\omega$.

**restoring force**
This is the force exerted on a body which is displaced from its equilibrium position (*see also* **spring force**).

**return map**
Let $V_n$ ($n \in I$) be a time series, where $I$ is a countable set. The map in which the $(n+1)$th point is plotted as a function of the preceding $n$th point $V_n$ is called a return map (or **Poincaré map**).

The construction of a return map is visualised in Figure 105 where an orbit in phase space is investigated at each crossing of a plane.

*Reference*: Lorenz (1963).

**reverse bifurcation** (*see* **bifurcation**).

**Reynolds number**
This is a dimensionless parameter used in describing the characteristics of a viscous fluid in motion and is given by $R = UL/T$, where $U$ is the flow speed, $L$ is a characteristic length scale of the flow, and $T$ is the kinematic viscosity.

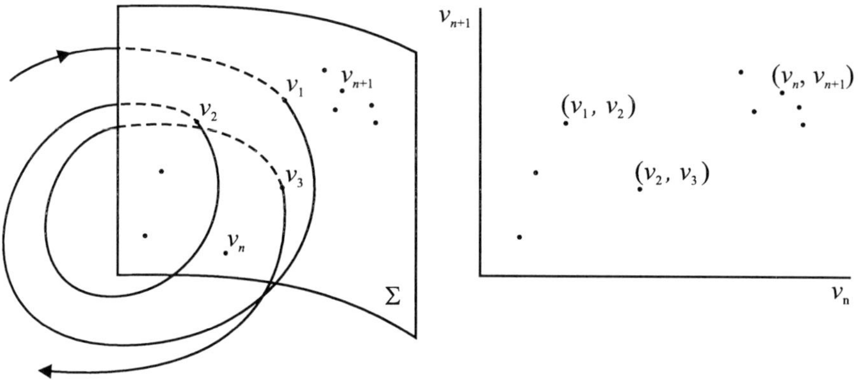


**Figure 105** Construction of a return map

Typically, a large Reynolds number suggests that viscous effects are negligible, but often at high Reynolds numbers steady flow is usually unstable so that the most likely visible motion is turbulent. On the other hand, a low Reynolds number suggests that the fluid is highly viscous, often resulting in laminar flow.

**Riccati equation**
The first-order ordinary differential equation

$$\frac{\mathrm{d}x}{\mathrm{d}t} = a_0(t) + a_1(t)x + a_2(t)x^2$$

where $x \in \mathcal{R}$ and $a_{0,1,2}$ are continuous bounded functions of time $t$, is called the Riccati equation.

**Richardson model**
This is a set of logistic equations which were developed by L.F. Richardson in the 1960s to model the arms race between two nations. In discrete form, these are given by

$$x_{n+1} = 4ay_n(1 - y_n)$$

and

$$y_{n+1} = 4bx_n(1 - x_n)$$

where $x_n$ and $y_n$ and $a, b$ are constants. The variables $x_n$, and $y_n$, denote the fractions of available resources that two countries devote to agreement.

*Reference*: Alligood *et al.* (1997).

**Ricker map**
This is the one-dimensional map

$$x_{n+1} = x_n \exp[r(1 - x_n)]$$

where $r$ is a constant, which was used originally by W.E. Ricker to model the dynamics of populations of fish species.

**riddled basins**
A riddled basin is one for which every neighbourhood of a point in the basin contains a portion of another attractor's basin. If $r_0$ is any point in a riddled basin of an attractor, then a ball in phase space of radius $\epsilon$ about $r_0$ has non-zero fraction of its volume in some other attractor's basin. Furthermore, this holds true no matter how small the value of $\epsilon$.

If the second basin is also riddled by the first basin, then we say that the basins are intermingled.

*Remark 1*: There is always a positive probability that an arbitrarily small perturbation in $r_0$ can move an initial condition in a riddled basin to another attractor.

An attractor $A$, whose basin $b(A)$ has a positive Lebesgue measure, has a locally riddled basin if there exists an $\epsilon > 0$ such that for every point $x \in b(A)$ any arbitrary small ball centred on $x$ contains a positive measure set of points whose orbits exceed a distance $\epsilon$ from $A$.

*Remark 2*: This second definition generalises the previous one to include the possibility that $b(A)$ contains an open neighbourhood of $A$.

The differences between a fractal basin boundary, and riddled and intermingled basins, are visualised in Figure 106. In Figure 106(a), basins of attractor $A_-$ (white) and attractor $A_+$ (black) have fractal basin boundary. The basin of attractor $A$ is riddled by basins of attractor $B$ in Figure 106(b), while basins of attractors $A$ and $B$ are intermingled in Figure 106(c).

*References*: Alexander *et al.* (1992); Ashwin *et al.* (1994).

**Rikitake model**
The Rikitake model describes the earth's magnetic field by the use of the following equations:

$$\frac{\mathrm{d}x_1}{\mathrm{d}t} = -vx_1 + yx_1$$

$$\frac{\mathrm{d}x_2}{\mathrm{d}t} = -vx_2 + (y - a)x_1$$

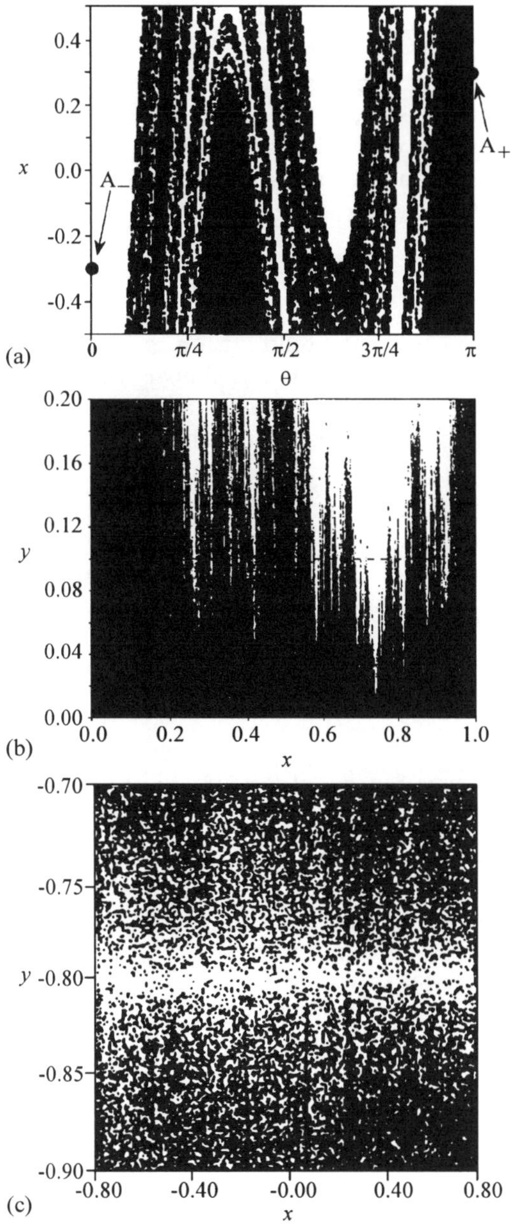


**Figure 106** Illustration of the differences between (a) fractal basin boundaries, (b) riddled basins and (c) intermingled basins; the basin of attractor $A$ (line at $y = 0$) is riddled by basins of attractor $B$ in (b), while basins of attractors $A$ and $B$ are intermingled in (c) ($A$ and $B$ indicated in black and white, respectively)

and

$$\frac{\mathrm{d}y}{\mathrm{d}t} = 1 - x_1 x_2$$

where $x_1$ and $x_2$ are the currents in two eddies in the earth's core, $y$ is the angular velocity of one eddy, $a > 0$ is the constant difference between the angular velocities of the two eddies, and $r$ is the ratio of the mechanical time scale to the magneto-diffusion time scale.

*Reference*: Drazin (1992).

**RLC circuit**
An RLC circuit is an electrical circuit with a resistor $R$, an inductor $L$ and a capacitor $C$. Currently, the most cited nonlinear circuit is the Chua or double-scroll circuit. Mathematical models consisting of a set of ordinary differential equations can be derived to describe the behaviour of the circuit (*see* **Chua's circuit**).

**robust**
Models, or solutions, that are insensitive to small perturbations in the governing parameters are said to be robust (*see also* **structural stability**).

**rotation number** (*see* **winding number**)

**routes to chaos**
There are basically four ways in which a system can become chaotic as a result of a change in parameters:

(i) period-doubling (Feigenbaum);

(ii) intermittency (Pomeau and Manneville);

(iii) subcritical instability;

(iv) a sequence of global bifurcations (Ruelle–Takens–Newhouse).

*Reference*: Ott (1992).

**Rössler equations**
The ordinary differential equations

$$\frac{\mathrm{d}x}{\mathrm{d}t} = -y - z$$

$$\frac{\mathrm{d}y}{\mathrm{d}t} = x + ay$$

and

$$\frac{\mathrm{d}z}{\mathrm{d}t} = b + z(x - c)$$

where a,b and $c$ are constants, are called the Rössler equations.

*Example*: For $a = b = 1/5$, and $c = 4.0$ and 5.7, the Rössler equations have, respectively, the periodic and chaotic attractors shown in Figure 107.

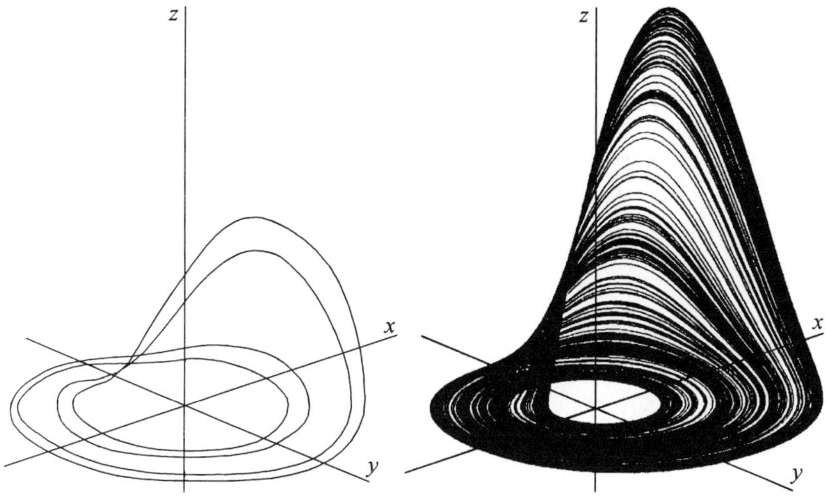


**Figure 107** (a) Periodic and (b) chaotic attractor of the Rössler system

*Reference*: Rössler (1976).

**Runge–Kutta method**

Numerical solutions of differential equations and difference equations are important tools in the study of nonlinear systems. Most of the nonlinear differential equations which can be found in practical applications are difficult, if not impossible, to solve analytically. In these cases, we typically rely on the use of numerical methods.

Among the many successful codes designed to integrate initial-value problems of ordinary differential equations, Runge–Kutta-type methods appear to be the most popular. Combined with the error formulae of Fehlberg, these methods have proved to be robust and widely applicable. We focus on the non-stiff situation and present the formulas on which a Runge–Kutta–Fehlberg method of order four is based. Suppose that the initial-value problem

$$\frac{\mathrm{d}u}{\mathrm{d}t} = f(t, u), \qquad u(t_0) = u_0$$

where $u \in \mathcal{R}^n$, is to be integrated. A typical integration step approximates $u$ at $t = t_0 + \Delta$, where $\Delta$ is the step length. The formulae are

$$u = u_0 + \Delta \sum_{k=0}^{4} c_k f^{(k)}$$

$$\bar{u} = u_0 + \Delta \sum_{k=0}^{5} \bar{c}_k f^{(k)}$$

with

$$f^{(0)} = f(t_0, u_0)$$

and

$$f^{(k)} = f(t_0 + \alpha_k \Delta, u_0 + \Delta \sum_{j=0}^{k-1} \beta_{kj} f^{(j)}).$$

TABLE 1

| $k$ | $\alpha_k$ | $\beta_{k0}$ | $\beta_{k1}$ | $\beta_{k2}$ | $\beta_{k3}$ | $\beta_{k4}$ | $c_k$ | $\bar{c}_k$ |
|---|---|---|---|---|---|---|---|---|
| 0 | 0 | 0 | | | | | $\frac{25}{216}$ | $\frac{16}{135}$ |
| 1 | $\frac{1}{4}$ | $\frac{1}{4}$ | | | | | 0 | 0 |
| 2 | $\frac{3}{8}$ | $\frac{3}{32}$ | $\frac{9}{32}$ | | | | $\frac{1408}{2565}$ | $\frac{6656}{12825}$ |
| 3 | $\frac{12}{13}$ | $\frac{1932}{2197}$ | $-\frac{7200}{2197}$ | $\frac{7296}{2197}$ | | | $\frac{2197}{4104}$ | $\frac{28561}{56430}$ |
| 4 | 1 | $\frac{439}{216}$ | -8 | $\frac{3680}{513}$ | $-\frac{845}{4104}$ | | $-\frac{1}{5}$ | $-\frac{9}{50}$ |
| 5 | $\frac{1}{2}$ | $-\frac{8}{27}$ | 2 | $-\frac{3544}{2565}$ | $\frac{1859}{4104}$ | $-\frac{11}{40}$ | | $\frac{2}{55}$ |

Both $u$ and $\bar{u}$ approximate the exact solution, with $\bar{u}$ being the approximation of higher order. The difference $u - \bar{u}$ serves as an estimate of the error in $u$.

*Reference*: Press *et al.* (1986).

**Ruelle inequality**
The inequality is defined as

$$h_\mu \leq \sum_{\lambda_j < 0} \lambda_j$$

where $\{\lambda_j\}$ are Lyapunov exponents, and $h_\mu$ is the measure theoretic entropy.

**Ruelle–Takens route to chaos**
This concept was first proposed by D. Ruelle and F. Takens in 1971, and latter modified with the assistance of S. Newhouse (hence it is sometimes also referred to as the Ruelle–Takens–Newhouse route to chaos), as a transition to turbulent motion in a fluid. A quasi-periodic solution with more than three frequencies is in general unstable, so that after only a few bifurcations turbulence occurs with the break up of a homoclinic orbit. (Compare this with the period-doubling route in which chaos occurs after an infinity of bifurcations.)

The typical scenario is thus: steady state $\rightarrow$ periodic $(\omega_1)$ $\rightarrow$ quasi-periodic $(\omega_1, \omega_2)$ $\rightarrow$ chaos.

*Reference*: Ruelle and Takens (1971).

# S

**saddle** (*see* **fixed point**)

**saddle connection** (*see* **global bifurcations**)

**saddle–node bifurcation** (*see* **bifurcation**)

**saddle point** (*see* **hill-top saddle**)

**safe (continuous) bifurcation**
This is a bifurcation in which the attracting set is continuous in control-parameter phase space, as the parameter goes through the bifurcation point, e.g. a supercritical Hopf bifurcation.

*Reference*: Thompson and Stewart (1986).

**Sarkovskii (Sharkovsky) theorem**
Let $T \subset \mathcal{N}$ be the ordered set $\{3 \prec 5 \prec 7 \prec ... \prec 2 \times 3 \prec 2 \times 5 \prec 2 \times 7 \prec ... \prec 2^2 \times 3 \prec 2^2 \times 5 \prec 2^2 \times 7 \prec ... \prec 8 \prec 4 \prec 2 \prec 1\}$. Let $f : I \to I = [0, 1]$, be a smooth map such that $f(0) = f(1) = 0$, which has a single fixed point. If $m < n$ relative to the order in the set $T$, and $f$ has a periodic point with a prime (i.e., shortest) period $m$, then $f$ has a periodic point of period $n$ (*see also* **Li–Yorke chaos**).

*References*: Devaney (1989); Sharkovsky (1964).

**sawtooth map** (*see* **Bernoulli shift map**)

**Schrödinger equation**
The partial differential equation

$$i\frac{\partial \phi(x,t)}{\partial t} + \frac{\partial^2 \phi(x,t)}{\partial x^2} + \phi(x,t)|\phi(x,t)|^2 = 0$$

where $i = \sqrt{-1}$, $x \in \mathcal{R}$ and $t \in \mathcal{R}^+$, is called the nonlinear Schrödinger equation.

**Schwarzian derivative**
This derivative, $Sf(x)$, of a smooth function $f(x)$ is given by

$$Sf(x) = \frac{\mathrm{d}^3 f/\mathrm{d}x^3}{\mathrm{d}f/\mathrm{d}x} - \frac{3}{2}\left(\frac{\mathrm{d}^2 f}{\mathrm{d}x^2}\frac{\mathrm{d}f}{\mathrm{d}x}\right)^2 .$$

*Remark*: The Schwarzian derivative has the following properties:

(i) Suppose that $Sf < 0$ and $Sg < 0$, then $S(fg) < 0$.

(ii) Suppose that $Sf < 0$ and $f$ has $n$ critical points, then $f$ has at most $(n+2)$ attracting periodic points.

The Schwarzian derivative is used in the theory of S-unimodal maps.

*Example*: Consider the function $f(x) = 4x(1 - x^2)$, the Schwarzian derivative is then given by

$$Sf(x) = -\frac{3}{4}\left(\frac{2}{2x-1}\right)^2 .$$

*Reference*: Devaney (1989).

**secular terms** (*see* **small parameter methods**)

**secure communication**
It has been proposed that the ability of coupled chaotic systems to synchronise may be used to mask the information of an input signal $I(t)$ by adding it to a larger chaotic signal $n(t)$ and transmitting the superposition of both signals. Information can be recovered after comparison of the received signal $I(t)+n(t)$ with the original chaotic noise $n(t)$. In this procedure, chaotic signals in the transmitter and receiver systems must be synchronised. As this way of sending information is difficult to unmask and hence is called secure communication. The main idea of secure communication is illustrated in Figure 108.

*Reference*: Cuomo and Oppenheim (1993).

**self-exciting system**
This term means that the system has no external source of energy upon which it can draw. Self-excited system is also sometimes called autocatalytic system.

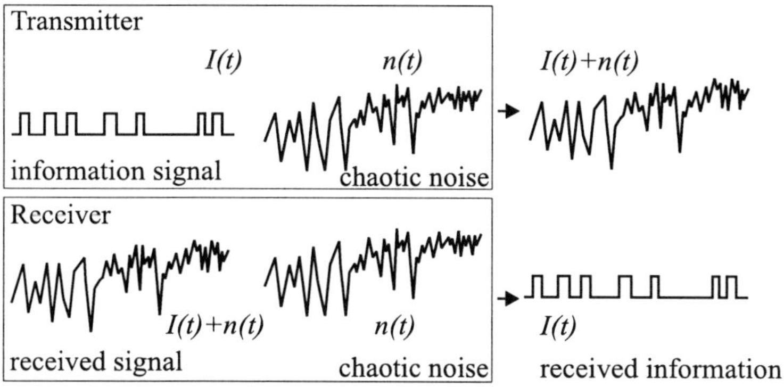


**Figure 108** The idea of secure communication

*Example*: Consider the Rayleigh equation

$$m\frac{\mathrm{d}^2x}{\mathrm{d}t}+\left[-a+b\left(\frac{\mathrm{d}x}{\mathrm{d}t}\right)^2\right]\frac{\mathrm{d}x}{\mathrm{d}t}+kx=0$$

with $a, b > 0$. The term

$$\left[-a+b\left(\frac{\mathrm{d}x}{\mathrm{d}t}\right)^2\right]\frac{\mathrm{d}x}{\mathrm{d}t}$$

means that for small $\mathrm{d}x/\mathrm{d}t$ the dissipation is negative and energy is injected into the system. The fixed point $(0,0)$ is unstable, so that with any perturbation from this point the system is self-excited, and the system remains bounded only due to the presence of the term $b(\mathrm{d}x/\mathrm{d}t)^3$.

**self-organised criticality**
This is a mechanism proposed to explain the physics of fractals. A spatially extended dynamical system evolves spontaneously into barely stable structures of critical states at which the spatial features of the system have self-similarity. This self-organised criticality is the common underlying mechanism for many self-similar and fractal phenomena. It is primarily studied in models of cellular automata.

*Reference*: Bak *et al.* (1987).

**self-similarity**
The property of an object or set of points such that the geometric structure is repeated at different scales or magnifications is known as self-similarity (*see also* **fractal sets**).

**sensitive dependence on initial conditions**

Consider a map $f(x) : \mathcal{R}^n \to \mathcal{R}^n$. Let $x_0 \approx x_0'$ be two slightly different, arbitrarily chosen, initial conditions and consider the two orbits

$$x_0, f(x_0), ..., f^n(x_0), ...$$

and

$$x_0', f(x_0'), ..., f^n(x_0'), ... .$$

If, after a finite number of iterations of the map, these two orbits become completely unrelated, then we have sensitive dependence on initial conditions (*see* Figure 109).

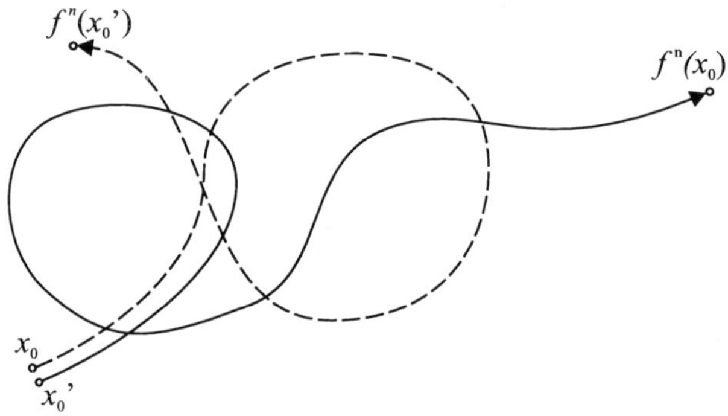


**Figure 109** Illustration of sensitive dependence on initial conditions

Sensitive dependence is a characteristic property of all chaotic systems, either maps or flows. The term 'butterfly effect' is sometimes colloquially used since this behaviour was first noticed by Lorenz in a paper entitled "Does the flap of a butterfly's wings in Brazil set off a tornado in Texas?".

A measure of the sensitivity can be found from calculating the divergence properties of the system by determining the Lyapunov exponents which measure the exponential divergence of close-by trajectories.

*Reference*: Lorenz (1993).

**separatrix**

Let $x \in \mathcal{R}^2$ be a saddle fixed point. An orbit which converges to the saddle point as $t \to \infty$ is called the stable manifold, inset or incoming separatrix, whereas if it converges to the saddle point as $t \to -\infty$, it is called then unstable manifold, outset or outgoing separatrix.

*Remark*: In single-degree-of-freedom driven oscillators the inset forms the separatrix between adjacent basins of attraction.

**shadowing lemma**
Let $A$ be a hyperbolic invariant set. Then for every $\beta > 0$, there is an $\alpha > 0$ such that every $\alpha$-pseudo orbit $\{x_i\}_{i=a}^{b}$, where $a \leq i < b$ in $A$ is $\beta$-shadowed by a point $y \in A$.

*Remark*: This lemma is proved only on uniformly hyperbolic sets.

*Reference*: Guckenheimer and Holmes (1993).

**shadowing orbit** (*see* **$\alpha$-pseudo orbit**)

**Sharkovsky's theorem** (*see* **Sarkovskii theorem**)

**shear map**
Let $x \in \mathcal{R}^2$ and $F : \mathcal{R}^2 \to \mathcal{R}^2$, and $f(x)$ be a differentiable function, and consider the shift $S(x) = (x, y + f(x))$, together with the rotation $R(x) = (y, -x)$, where $S, R : \mathcal{R}^2 \to \mathcal{R}^2$; then the map

$$f(x) = S(R(S(x)))$$

is called a shear map.

**shift operator**
This operator, $\mathcal{S}$, acts on the symbol sequences $\Sigma = s_0 s_1 s_2 ... s_k s_{k+1} ...$, etc. in the following way:

$$\mathcal{S}(\Sigma) = s_1 s_2 ... s_k s_{k+1} ...$$

and

$$\mathcal{S}^k(\Sigma) = s_k s_{k+1} ....$$

(*see also* **symbolic dynamics**).

**Shilnikov theorem**
Consider a flow $\phi_t$ in $\mathcal{R}^3$, which has a fixed point at the origin with a real eigenvalue $\hat{\lambda} > 0$, and a pair of complex eigenvalues $\omega$ and $\bar{\omega} \neq \omega$, which have negative real parts. We introduce coordinates in such a way that the linearised version of the stable manifold $W^s$, i.e. the stable subspace, is contained in the $(x, y)$ plane and the unstable manifold $W^u$ contains the $z$-axis. Assume that the trajectory $\gamma$ in $W^u(0)$ which points upwards near 0 is a homoclinic trajectory which enters the $(x, y)$ plane and spirals towards the origin as $t \to \infty$, as shown in Figure 110.

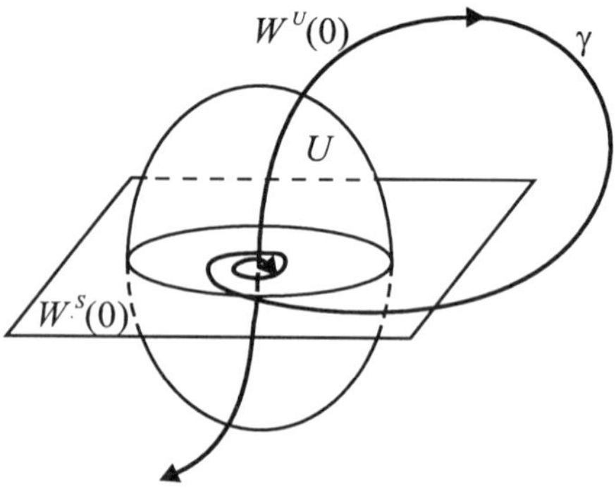


**Figure 110** A homoclinic trajectory to a saddle focus

If $|\text{Re}\ \omega| < \hat{\lambda}$, then the flow $\phi_t$ can be perturbed to $\phi'$, such that $\phi'$ has a homoclinic orbit $\gamma'$ near $\gamma$ and the return map of $\gamma'$ for $\phi'_t$ has a countable set of horseshoes.

*References*: Guckenheimer and Holmes (1983); Wiggins (1990).

**shooting method**
This is a method which applied to the numerical solution of continuum systems. Through an iterative scheme, adjustments are made to the parameters of a numerically computed solution to an initial-value problem in order to satisfy far-end boundary conditions.

*Reference*: Press *et al.* (1986).

**Sierpinski carpet**
A Sierpinski carpet is a fractal formed when a central cross or square is removed from an initial square and the process repeated on each of the remaining smaller squares (as shown in Figure 111).

**Sierpinski gasket**
Consider the following construction (shown in Figure 112). The first step is a filled triangle. In the second step, the central triangle is eliminated. The next step eliminates all of the central triangles from each filled triangle. When the number of steps increases to infinity we then obtain a fractal set which is called the Sierpinski gasket.

*Remark*: The capacity dimension of the Sierpinski gasket is

$$D = \ln 3 / \ln 2 = 1.58.$$

*Reference*: Barnsley (1988).

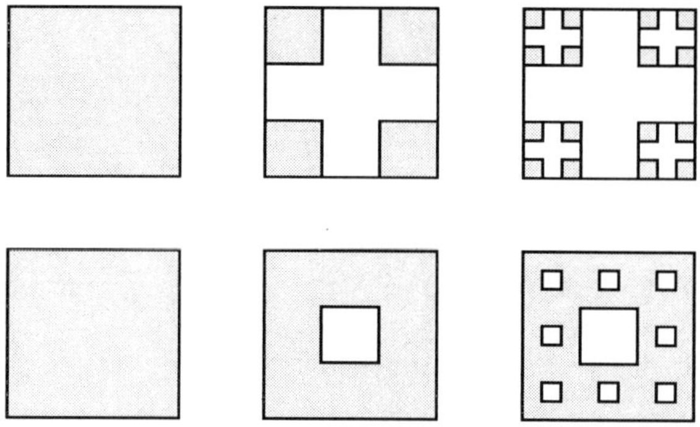

**Figure 111** Steps in the construction of a Sierpinski carpet

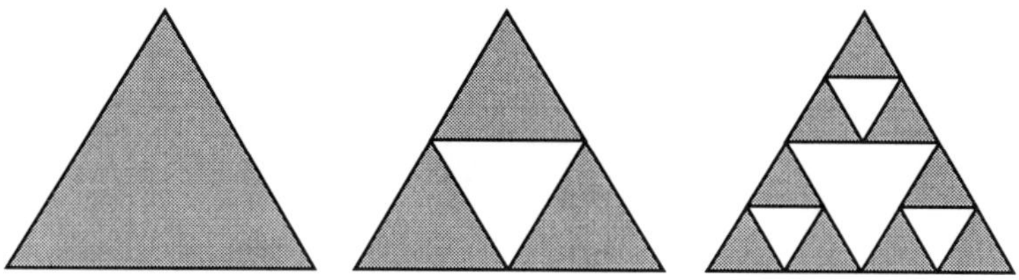

**Figure 112** Steps in construction of a Sierpinski gasket

**$\sigma$-algebra (sigma algebra)**
Let $X$ be a set. A $\sigma$-algebra of subsets of $X$ is a collection $\mathcal{B}$ of subsets $B$ of $X$ satisfying the following conditions:

$$X \in \mathcal{B}$$

$$B \in \mathcal{B} \Rightarrow X \backslash B \in \mathcal{B}$$

and

$$B_n \in \mathcal{B} \Rightarrow \cup_{n=1}^{\infty} B_n \in \mathcal{B}.$$

*Reference*: Walters (1982).

**simple harmonic motion**
This is the periodic oscillatory motion governed by the linear equation

$$\frac{\mathrm{d}^2 x}{\mathrm{d}t^2} + kx = 0$$

where $k > 0$ is constant.

**Sinai–Ruelle–Bowen (SRB) measure** (*see* **natural measure**)

**sine–Gordon equation**
The partial differential equation

$$\frac{\partial^2 \phi(x,t)}{\partial x^2} - \frac{\partial^2 \phi(x,t)}{\partial t^2} = \sin \phi(x,t)$$

where $x \in \mathcal{R}$, is called the sine–Gordon equation.

**sine map**
This map is a one-dimensional unimodal map

$$x_{n+1} = r \sin \pi x_n$$

where $0 \leq r \leq 1$, $0 \leq x \leq 1$, which has a bifurcational structure similar to that of the logistic map.

**Singer's theorem**
This is a theorem for one-dimensional maps which states the following:

Let $f : I \to I$ be a $C^3$ map with negative Schwarzian derivative. If $\gamma$ is a stable periodic orbit, then there exists a critical point $x^*$, such that $f'(x^*)$, or an endpoint of $I$ whose trajectory approaches $\gamma$.

*Reference*: Thompson and Stewart (1986).

**singular points** (*see* **fixed points**)

**sink**
This is an asymptotically stable fixed point.

**skinny baker map**
The skinny baker map is a discontinuous map on $\mathcal{R}^2$ which contains stretching in one direction and shrinking in the other, given for example by

$$B(x,y) = \begin{cases} (\frac{1}{3}x, 2y) & \text{if} \quad 0 \leq y \leq \frac{1}{2} \\ (\frac{1}{3}x + \frac{2}{3}, 2y - 1) & \text{if} \quad \frac{1}{2} < y \leq 1 \end{cases}$$

(*see also* **baker map**).

*Reference*: Alligood *et al.* (1997).

**slaving principle**
This is the behaviour in which one subsystem (slave) follows the behaviour of an other subsystem (master). This phenomenon is similar to synchronisation.

The term slaving principle was notably used by Haken and applied to systems near singularities.

*Reference*: Haken (1976).

**slowly varying variables** (*see* **averaging methods**)

**Slutsky–Frisch–Tinbergen method**
This is a methodology in economic modelling method based on the use of exogenous stochastic impulses which are transformed into oscillation patterns through the filtering properties of the economy's linear 'propagation method'.

*Reference*: Grebogi and Yorke (1997).

**Smale–Birkhoff theorem**
This theorem, which is also known as the Smale–Birkhoff homoclinic theorem, establishes a criterion for the existence of hyperbolic invariant sets in a flow.

*Theorem*: Let $f : \mathcal{R}^n \to \mathcal{R}^n$ be a diffeomorphism such that $p$ is a hyperbolic fixed point and there exists a point $q \neq p$ of transversal intersection between the stable $W^s(p)$ and unstable $W^u(p)$ manifolds of $p$. Then $f$ has a hyperbolic invariant set $A$ on which $f$ is topologically equivalent to a subshift of finite type.

*Reference*: Guckenheimer and Holmes (1983).

**Smale horseshoe** (*see* **horseshoe map**)

**small-parameter methods**
This is a term used for perturbation methods. Consider the equation

$$\frac{\mathrm{d}^2 x}{\mathrm{d}t^2} + \omega_0^2 x + \epsilon f\left(x, \frac{\mathrm{d}x}{\mathrm{d}t}, t\right) = 0 \tag{1}$$

where $x \in \mathcal{R}$ and $\epsilon << 1$ is a small parameter. An approximate method for solving (1) assumes that the periodic solution has the form of a power series

$$x(t) = x_0(t) + \epsilon x_1(t) + \epsilon^2 x_2(t) + \ldots \tag{2}$$

and a frequency

$$\omega^2 = \omega_0^2 + \epsilon a_1 + \epsilon^2 a_2 + \ldots$$

where $a_{1,2,\ldots}$ are constants chosen in such a way so as to avoid secular terms of the type $t \sin \omega t$ in the solution (2). Methods of this type are called small-parameter methods.

These methods are very useful for estimating analytical solutions of weekly nonlinear systems.

*References*: Huntley and Johnson (1983); Nayfeh and Mook (1979).

**snowflake** (*see* **Koch curve**)

**soliton**
The term soliton describes a solitary, uniformly propagating disturbance, which preserves its structure and velocity after an interaction with another soliton. A soliton is a wave solution to a partial differential equation.

*Reference*: Drazin and Johnson (1989).

**solution curve**
This is another name for a trajectory or orbit.

**source**
This is an unstable fixed point with two positive eigenvalues.

**spatial–temporal chaos**
This term (for chaos in both space and time) describes a temporal (high-dimensional) chaos which involves spatial pattern dynamics. There is no strict mathematical definition but the term is used to cover a range of problems in which a suitably complex invariant manifold exists, such as elastica, Josephson junction arrays, turbulence in plasma, etc.

**spectral analysis**
Let the time evolution of a dynamical system be represented by the time variation $x(t)$ (time series) of its dynamical variables. In many practical cases the time-dependent function $x(t)$ could be represented as a superposition of its periodic components. The determination of these components is called spectral analysis.

If the function $x(t)$ is continuous and its derivative $\mathrm{d}f(t)/\mathrm{d}t$ is also continuous, and $x(t)$ is periodic, i.e.

$$x(t) = x(t + nT)$$

with $n$ being a positive or a negative integer and $T$ being the basic periodicity, then $x(t)$ can be expressed as a linear combination of oscillations whose frequencies are integer multiples of a basic frequency $\omega_0$ i.e.

$$x(t) = \sum_{n=-\infty}^{\infty} (a_n \cos n\omega_0 t + b_n \sin n\omega_0 t) \tag{1}$$

or by using complex notation:

$$x(t) = \sum_{n=-\infty}^{\infty} c_n e^{in\omega_0 t} \tag{2}$$

where $a_n$, $b_n$ and $c_n$ are constants. The functional series, (1) and (2), are called *Fourier series*. The amplitudes of the components of the frequency $n\omega_0$ are given by

$$a_n = \frac{\omega_0}{2\pi} \int_{-\pi/\omega_0}^{\pi/\omega_0} x(t) \cos(n\omega_0 t) \mathrm{d}t \tag{3$a$}$$

$$b_n = \frac{\omega_0}{2\pi} \int_{-\pi/\omega_0}^{\pi/\omega_0} x(t) \sin(n\omega_0 t) \mathrm{d}t \tag{3$b$}$$

or

$$c_n = \frac{\omega_0}{2\pi} \int_{-\pi\omega_0}^{\pi/\omega_0} x(t) e^{-in\omega_0 t} \mathrm{d}t. \tag{3$c$}$$

*Reference*: Kaplan (1973).

**spiral**
This is a different name for a focus (*see* **fixed points**).

**spring force**
Many physical systems can be modelled by an equivalent spring/mass system in which the spring exerts a force $k(x)$ on a mass $m$ (as shown in Figure 113(a)).

The function $k(x)$ is called a spring characteristic. For a linear spring,

$$k(x) = Kx,$$

where $K$ is a spring stiffness constant coefficient. For a nonlinear spring, $k(x)$ is nonlinear, and we say that for a hard spring $k(x)$ increases with displacement, while if $k(x)$ decreases then $k(x)$ is said to be a soft spring (see Figure 113(b)).

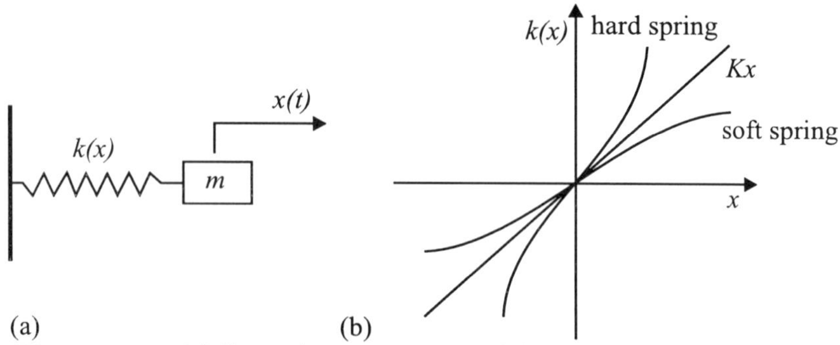


**Figure 113** (a) Spring/mass model and (b) different characteristics of a spring

**square root singularity**
In simple models of impact oscillators an approximation to the Poincaré map, close to the grazing point where impacts first occur, includes a square root singularity. The approximation, following A. B. Nordmark, is given by

$$x_{n+1} = \alpha x_n + y_n + \rho$$

and

$$y_{n+1} = -4x_n$$

for $x_n \leq 0$, and for $x_n > 0$ by

$$x_{n+1} = -\sqrt{x_n} + y_n + \rho$$

and

$$y_{n+1} = -\gamma\tau^2 x_n.$$

In the above, $\tau^2$ is a constant related to the coefficient of restitution, and typically $\gamma$ and $\alpha$ are fixed constants with $\rho$ varying through zero at the point of grazing.

A similar one-dimensional map has the form

$$x_{n+1} = \sqrt{d - x_n} + rx_n$$

for $x \leq d$, and

$$x_{n+1} = rx_n$$

for $x > d$, where $r$ is a restitution coefficient.

*References*: Foale and Bishop (1994); Nordmark (1991).

**squid axon** (*see* **axon**)

**SRB measure** (*see* **natural measure**)

**stability**

Consider any particular solution (say $x = u(t)$) of the equation

$$\frac{\mathrm{d}x}{\mathrm{d}t} = f(x, t, c)$$

where $x \in \mathcal{R}^n$, and $c \in \mathcal{R}^m$. An important question is whether this solution is stable, i.e. will small perturbations cause the system to evolve away from this solution?

There are several definitions of stability. However, the most used commonly are the following:

*Stability in the sense of Lyapunov*

(i) The solution $u(t)$ is said to be uniformly stable if there exists a $\delta(\epsilon) > 0$ for every $\epsilon > 0$, such that any other solution $v(t)$, for which $|u - v| < \delta(\epsilon)$ at $t = t_0$, satisfies $|u(t) - v(t)| < \epsilon$ for all $t \geq t_0$. If no such $\delta(\epsilon)$ exists, then $u(t)$ is said to be unstable.

(ii) If $u(t)$ is uniformly stable, and in addition

$$\lim_{t\to\infty} |u(t) - v(t)| \to 0$$

then $u(t)$ is said to be asymptotically stable.

*Remark*: These stability criteria are quite restrictive as they require that $u(t)$ and $v(t)$ remain close to each other for the same values of time $t$ in both solutions (*see* time 'ticks' in Figure 114).

*Stability in the sense of Poincaré*

(i) Let $\Gamma$ be the orbit defined by $u(t)$ for all $t$, and $\Gamma'$ be the orbit defined by the solution $v(t)$ for all $t$. We say that $\Gamma$ is orbitally stable if, for any $\epsilon > 0$, there exists a $\delta(\epsilon)$ such that, if $|u(0) - v(\tau)| < \delta(\epsilon)$ for some $\tau$, then there exists a $t'(t)$ such that $|u(t) - v(t')| < \epsilon$ for all $t > 0$.

(ii) The orbit $\Gamma$ is said to be asymptotically stable if $\Gamma'$ tends towards $\Gamma$ as $t \to \infty$.

*Remark*: Orbital stability implies that the two solutions will follow the same evolution but possibly on different time scales, related by $t'(t)$ (*see* Figure 115 – note here the difference in density of the 'time ticks' on the two orbits).

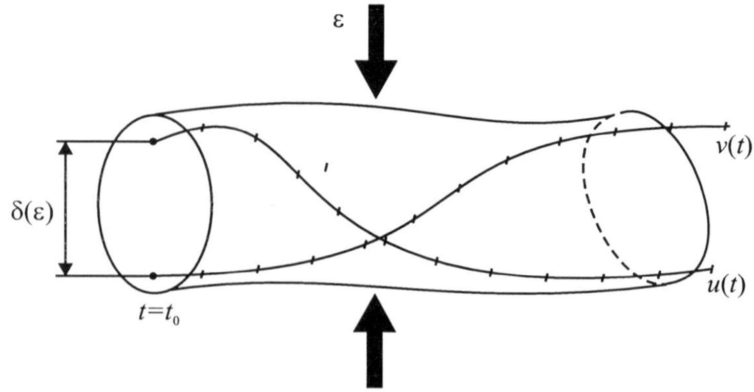


**Figure 114** Illustration of stability in the sense of Lyapunov

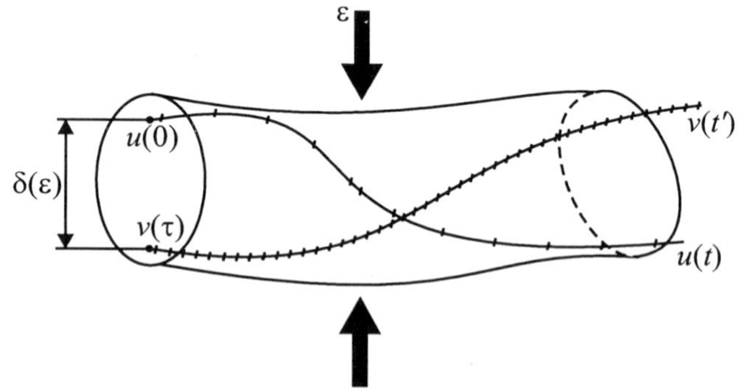


**Figure 115** Illustration of orbital stability

*Stability in the sense of Lagrange*

The solutions of

$$\frac{\mathrm{d}x}{\mathrm{d}t} = f(x, t)$$

are said to be stable in the sense of Lagrange if $M < \infty$, and $|x| \leq M$ for all $t$.

*Reference*: Jackson (1990).

**stable and unstable manifolds**
Let $\Phi_t(u_0)$ be a flow of the dynamical system

$$\frac{\mathrm{d}u}{\mathrm{d}t} = f(u)$$

where $u \in \mathcal{R}^n$, and $f(0) = 0$. A set $S$ is called a stable manifold of the critical point $u = 0$ if for all initial conditions $u_0 \in S$

$$\lim_{t \to \infty} \Phi_t(u_0) = 0.$$

In the same way, a set $U$ is called an unstable manifold of a critical point $u = 0$ if for all initial conditions $u_0 \in U$

$$\lim_{t \to -\infty} \Phi_t(u_0) = 0.$$

*Example*: Consider the Helmholtz oscillator

$$\frac{\mathrm{d}^2 u}{\mathrm{d}t^2} + \frac{\mathrm{d}u}{\mathrm{d}t} - u - u^2 = 0.$$

Let $u_1 = u$, and $u_2 = \mathrm{d}u/\mathrm{d}t$. Then we find the two fixed points

$$(u_1^*, u_2^*) = (0, 0), \qquad (u_1^*, u_2^*) = (-1, 0).$$

The linear analysis shows that the first of these fixed points is a saddle, while the second is a centre (Figure 116).

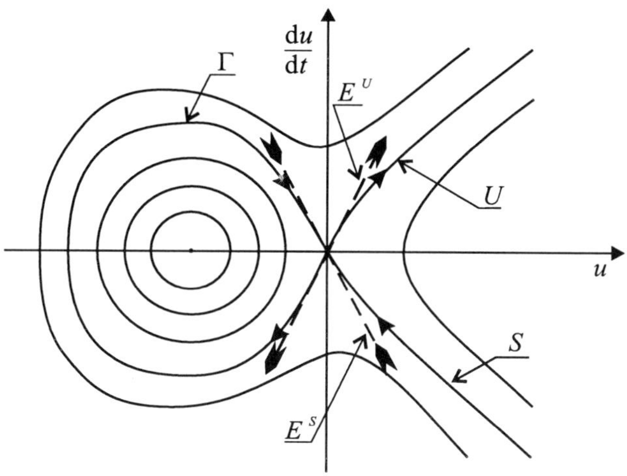


**Figure 116** Phase-space portrait of a Helmholtz oscillator

The total energy of the system $H$ (the sum of the kinetic energy, $(u/2)^2/2$, and the potential energy, $-u^2/2 - u^3/3$) is equal to

$$H = \frac{1}{2}\left(\frac{\mathrm{d}u}{\mathrm{d}t}\right)^2 - \frac{1}{2}u^2 - \frac{1}{3}u^3.$$

The solution curves in phase space are given by

$$\left(\frac{du}{dt}\right)^2 - u^2 - \frac{2}{3}u^3 = C.$$

At the saddle point, the stable and unstable subspaces of a linearised system $E^s$, $E^u$ are tangent to the stable and unstable manifolds of a nonlinear system $S, U$.

*Remark*: The stable manifold is sometimes referred to as the inset while the term outset used for the unstable manifold.

*Reference*: Thompson and Stewart (1986).

**stable and unstable subspaces**

Suppose that the $(n \times n)$-dimensional matrix $A$ has $k$ negative eigenvalues $\hat{\lambda}_1, \ldots, \hat{\lambda}_k$, and $(n-k)$ positive eigenvalues, $\hat{\lambda}_{k+1}, \ldots, \hat{\lambda}_n$, and that these eigenvalues are distinct. Let $\{\mathbf{v}_1, \ldots, \mathbf{v}_n\}$ be a corresponding set of eigenvectors. Then the stable and unstable subspaces of the linear system $dx/dt = Ax$, namely $E^s$ and $E^u$, are the linear subspaces spanned by $\{\mathbf{v}_1, \ldots, \mathbf{v}_k\}$ and $\{\mathbf{v}_{k+1}, \ldots, \mathbf{v}_n\}$, respectively:

$$E^s = \text{span}\{\mathbf{v}_1, \ldots, \mathbf{v}_k\}$$

and

$$E^u = \text{span}\{\mathbf{v}_{k+1}, \ldots, \mathbf{v}_n\}.$$

*Example*: Consider the linear system

$$\frac{du_1}{dt} = -u_1 - 3u_2$$

and

$$\frac{du_2}{dt} = 2u_2\,. \tag{1}$$

The global phase portrait of Figure 117 can be found by drawing the solution curves defined by

$$u_1(t) = c_1 e^{-t} + c_2(e^{-t} - e^{2t}),$$

and

$$u_2(t) = c_2 e^{2t}.$$

The arrows in the Figure 117 indicate the evolution of the system in time, while the lines through the origin are the stable and unstable subspaces of (1).

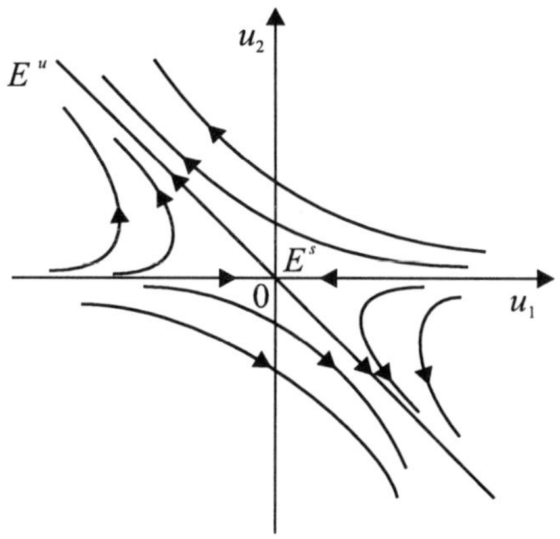


**Figure 117** Phase portrait for linear system near the saddle

Note that solutions starting on the stable subspace $E^s$ approach the critical point at origin as $t \to \infty$, and that solutions starting on the unstable subspace $E^u$ approach the critical point as $t \to -\infty$.

**standard map**
The map of the plane given by

$$S_0(x, y) = (x + y \quad (mod\ 2\pi); y + a\sin(x + y) \quad (mod\ 2\pi))$$

where $x, y \in [-\pi, \pi]$ and $a$ is constant, is called the standard map. $S_0$ has a toroidal phase space (this map is similar to Arnold's cat map).

**stationary solution**
A solution of ordinary or partial differential equations which is invariant in time is called stationary.

**steady state**
A steady-state solution is a time-invariant motion or equilibrium.

**stick–slip systems**
The term stick–slip is used to describe the periodic motion induced by the friction between two surfaces, which alternates between a sliding action (slip) and a period in which there is no relative velocity (stick). In its simplest form, stick–slip is a self-excited response in an autonomous system.

*Example*: Consider the motion of a spring-mounted block of mass $m$ riding on a conveyor belt which is being driven at a constant speed $v$ (see Figure 118(a)).

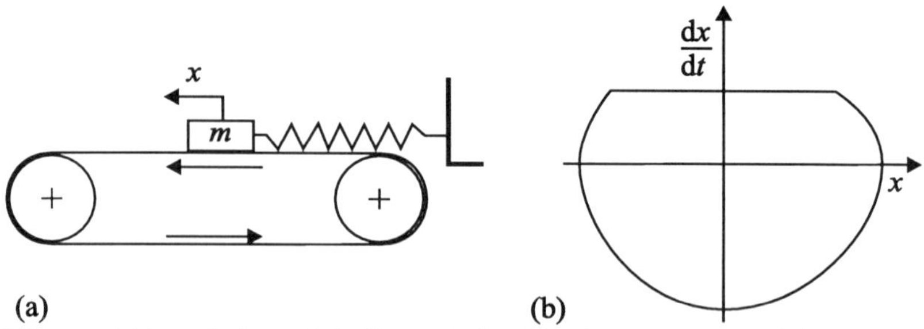


**Figure 118** (a) A model of a typical stick–slip system and (b) a phase portrait of stick–slip motion

A typical phase portrait of stick–slip motion is shown in Figure 118(b).

*Reference*: Popp and Stelter (1990).

**stochastic resonance**
This is the phenomenon in which the signal due to a weak periodic force in a nonlinear system is amplified by the addition of external random noise, thus producing a spike in the power spectrum.

*Reference*: Benzi *et al.* (1982).

**strange attractor**
An attractor which has a fractal structure in phase space (*see also* **attractor**) is called a strange attractor.

**strange nonchaotic attractor**
If an attractor has a fractal structure in phase space, but a typical trajectory on the attractor is characterised by non-positive Lyapunov exponents, then it is called a strange nonchaotic attractor.

*Example*: Consider the logistic map at the accumulation point (the accumulation of the period-doubling cascade) $a_\infty$. At this point, the Lyapunov exponent is zero but the dimension of the attractor is fractal, $d_c = 0.54$.

*Remark*: Strange nonchaotic attractors have been found to be common features in quasi-periodically forced systems.

*References*: Kapitaniak and Wojewoda (1993); Romeiras and Ott (1987).

**stretching and folding**
These are the actions which typify a chaotic system. Within the attractor,

close-by trajectories undergo exponential divergence as time evolves so that they are stretched and move apart, while globally the nonlinearity ensures that the trajectories are bounded by means of a folding action. This action is often compared to a baker kneading dough when making bread.

*Example*: For the logistic map, the stretching (divergence of nearby trajectories) and folding (confinement to bounded space), mechanisms are shown in Figure 119.

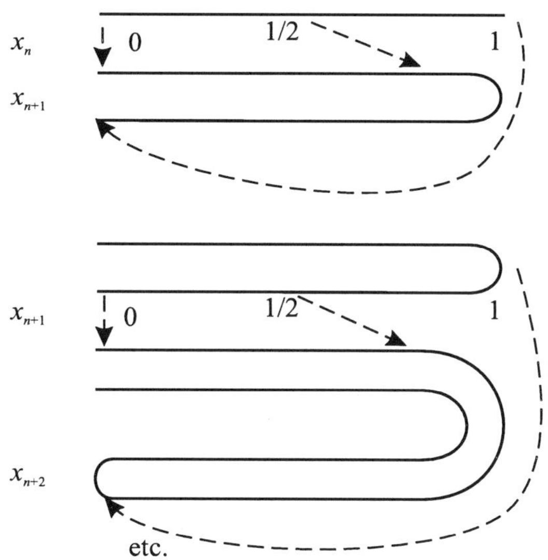


**Figure 119** The stretching and folding mechanism for the logistic map

**stroboscopic map**

A continuous system which is driven periodically may be inspected discretely at multiples of the driving period to produce a stroboscopic map (*see also* **Poincaré map**).

**strong resonances**

If we consider the way in which a two-dimensional map can become unstable we note that the saddle–node and period-doubling bifurcations are essentially one dimensional but that the flutter or Neimark bifurcation has a two-dimensional centre manifold with a pair of complex conjugate eigenvalues $\lambda$ crossing the unit circle. The special cases where $\lambda^n = 1$ ($n$=1–4) are called strong resonances.

**structural stability**
A dynamical system

$$\frac{\mathrm{d}x}{\mathrm{d}t} = f(x)$$

is structurally stable if its phase space is topologically equivalent to that of the system

$$\frac{\mathrm{d}x}{\mathrm{d}t} = f(x) + \delta f$$

where $\delta f$ is an arbitrary smooth $C^r$ function which is sufficiently small, i.e.

$$\max ||\delta f(x)|| \leq \epsilon$$

for all $x \in \mathcal{R}^n$, where $||.||$ is any norm in $\mathcal{R}^N$.

*Reference*: Peixoto (1977).

**subcritical bifurcation**
This is a bifurcation in which a pre-existing steady state becomes unstable (*see* **bifurcation**), usually resulting in the system evolving to a distant portion of the phase space.

**subharmonic resonance**
Consider the forced oscillator

$$\frac{\mathrm{d}^2x}{\mathrm{d}t^2} + f(x, \frac{\mathrm{d}x}{\mathrm{d}t}) = A\cos\Omega t.$$

If the driving frequency $\Omega$ and the oscillation frequency $\omega$ are related in the following way:

$$\omega = \frac{\Omega}{n}$$

where $n > 1$ is an integer, then the motion is subharmonic.

The resonance or large-amplitude response which occurs at

$$\Omega = n\omega$$

is called a superharmonic resonance.

**submanifold**
A submanifold $M$ of a manifold $N$ is a subset of $N$ which is a manifold.

**supercritical bifurcation**
This is a bifurcation in which a system changes its form but remains stable as a parameter is varied (*see* **bifurcation**).

**suspension of maps**
Typically the behaviour of closed orbits of a flow is evaluated by considering the equivalent behaviour of the associated fixed point of the Poincaré map. Conversely, for every diffeomorphism (a differential map whose inverse exists which is also differentiable) there exists a flow whose Poincáre map is precisely the same as the diffeomorphism. This flow is called a suspension of the map (diffeomorphism).

*Remark*: The topology of the suspended manifold may be complicated.

*Reference*: Arrowsmith and Place (1990).

**swallowtail**
One of the elementary catastrophes (*see* **Thom's theorem**).

**symbolic dynamics**
The representation of orbits of a map or a differential equation as symbol sequences is called symbolic dynamics.

**symmetry-breaking bifurcation**
This is a global bifurcation in which a symmetrical (with any type of symmetry) attractor is replaced by two co-existing asymmetrical attractors.

*Example*: A symmetrical chaotic attractor of the Lorenz equations shown in Figure 120(a) ($\sigma = 10, b = 8/3, andr = 200$) is destroyed by a symmetry-breaking bifurcation at $r = 203$ and replaced by two asymmetric attractors of the type shown in Figure 120(b).

**synchronisation of chaos**
An essential property of a chaotic trajectory is that it is not asymptotically stable. Closely correlated initial conditions have trajectories which quickly become uncorrelated. Despite this obvious disadvantage, it has been established that synchronisation of two chaotic systems is possible.

The basic synchronisation procedure can be described as follows. Suppose that an $n$-dimensional dynamical system

$$\frac{\mathrm{d}u}{\mathrm{d}t} = h(u)$$

can be divided into the two subsystems;

$$\frac{\mathrm{d}x}{\mathrm{d}t} = f(x, y)$$

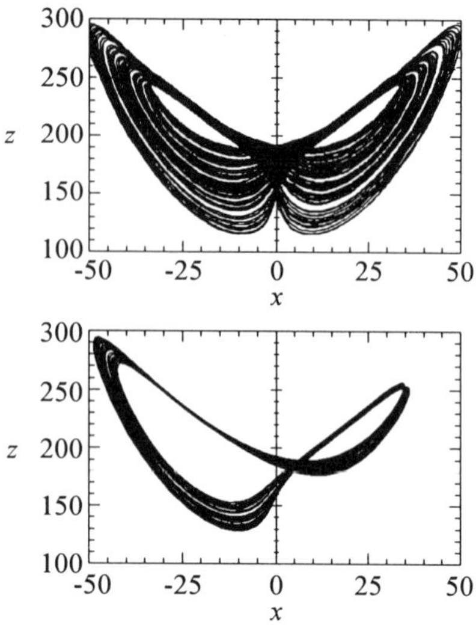


**Figure 120** (a) A symmetrical attractor and (b) one of the two asymmetrical attractors of the Lorenz equations

and

$$\frac{\mathrm{d}y}{\mathrm{d}t} = g(x, y) \tag{1}$$

where

$$u = (x, y)^T, \; x = (u_1, ..., u_m)^T, \; f = (h_1(u), ..., h_m(u))^T,$$
$$y = (u_{m+1}, ..., u_n)^T, \; g = (h_{m+1}(u), ..., h_n(u))^T.$$

Now create a new subsystem $z$ which is identical to the $y$ subsystem, substitute the set of variables $z$ for the corresponding variables $y$ in the function $g$, and augment equation (1) with this new system, thus giving

$$\frac{\mathrm{d}x}{\mathrm{d}t} = f(x, y)$$

$$\frac{\mathrm{d}y}{\mathrm{d}t} = g(x, y)$$

and

$$\frac{\mathrm{d}z}{\mathrm{d}t} = g(x, z). \tag{2}$$

The first two equations of (2) are called a *driving subsystem*, while the third equation is known as the *response subsystem*.

Lyapunov exponents of the response subsystem for a particular input $x(t)$ are called *conditional Lyapunov exponents*. Let $y(t)$ be a chaotic trajectory with initial condition $y(0)$, and $z(t)$ be a trajectory started at a different initial point $z(0)$. It has been shown that the necessary and sufficient condition for

$$|z(t) - y(t)| \to 0 \tag{3}$$

that is, for the two subsystems to be synchronised, is that all of the conditional Lyapunov exponents are negative.

We can describe this procedure by using an example of Chua's circuit (*see* **Chua's circuit**) in which the dimensionless equation can be decomposed in three different ways:

(i) An $x$-drive configuration where the state equation becomes

$$\frac{\mathrm{d}x}{\mathrm{d}t} = f(x,y), \qquad \frac{\mathrm{d}x}{\mathrm{d}t} = \alpha(y - x - f(x))$$

$$\frac{\mathrm{d}y}{\mathrm{d}t} = g(x,y), \qquad \frac{\mathrm{d}y}{\mathrm{d}t} = x - y + z$$

$$\frac{\mathrm{d}z}{\mathrm{d}t} = -\beta y$$

$$\frac{\mathrm{d}z}{\mathrm{d}t} = g(x,z), \qquad \frac{\mathrm{d}y'}{\mathrm{d}t} = x - y' + z'$$

and

$$\frac{\mathrm{d}z'}{\mathrm{d}t} = -\beta y'$$

(ii) A $y$ drive configuration with the state equations

$$\frac{\mathrm{d}x}{\mathrm{d}t} = f(x,y), \qquad \frac{\mathrm{d}y}{\mathrm{d}t} = x - y + z$$

$$\frac{\mathrm{d}y}{\mathrm{d}t} = g(x,y), \qquad \frac{\mathrm{d}x}{\mathrm{d}t} = \alpha(y - x - f(x))$$

$$\frac{\mathrm{d}z}{\mathrm{d}t} = -\beta y$$

$$\frac{\mathrm{d}z}{\mathrm{d}t} = g(x,z), \qquad \frac{\mathrm{d}z'}{\mathrm{d}t} = \alpha(y - x' - f(x'))$$

and

$$\frac{\mathrm{d}z'}{\mathrm{d}t} = -\beta y'$$

(iii) A $z$ drive configuration where the state equations are as follows

$$\frac{\mathrm{d}x}{\mathrm{d}t} = f(x,y), \qquad \frac{\mathrm{d}y}{\mathrm{d}t} = -\beta y$$

$$\frac{\mathrm{d}y}{\mathrm{d}t} = g(x,y), \qquad \frac{\mathrm{d}x}{\mathrm{d}t} = \alpha(y - x - f(x))$$

$$\frac{\mathrm{d}y}{\mathrm{d}t} = x - y + z$$

$$\frac{\mathrm{d}z}{\mathrm{d}t} = g(x,z), \qquad \frac{\mathrm{d}z'}{\mathrm{d}t} = \alpha(y' - x' - f(x'))$$

and

$$\frac{\mathrm{d}y'}{\mathrm{d}t} = x' - y' + z.$$

It can be shown that for $\alpha = 10$, $\beta = 14.87$, $a = -1.27$, and $b = -0.68$, the subsystems $\mathrm{d}y/\mathrm{d}t = g(x,y)$ and $\mathrm{d}z/\mathrm{d}t = g(x,z)$ can be synchronised only in the $x$- and $y$-drive configurations as the conditional Lyapunov exponents are $[\lambda_1^c = -0.05, \lambda_2^c = -0.05]$, $[\lambda_1^c = -2.55, \lambda_2^c = 0]$, and $[\lambda_1^c = -5.42, \lambda_2^c = 1.23]$, respectively, for the $x$, $y$ and $z$ configurations.

*References*: Kapitaniak (1996); Pecora and Carroll (1990).

**synergetics**

This is an interdisciplinary field of research concerned with the co-operation of individual parts of a system that produces macroscopic spatial, temporal or functional structures. This field covers deterministic as well as stochastic processes.

*Reference*: Haken (1976).

# T

**Takens–Bogdanov bifurcation**
Consider a planar system

$$\frac{\mathrm{d}x}{\mathrm{d}t} = f(x, \alpha)$$

where $x \in \mathcal{R}^2$, $\alpha \in \mathcal{R}^2$ and $f$ is smooth. Suppose that for $\alpha = 0$ this system has a fixed point $x = 0$ with two zero eigenvalues, and that the Jordan normal form of the Jacobian $\partial f / \partial x$ at the fixed point has a form

$$\begin{vmatrix} 0 & 1 \\ 0 & 0 \end{vmatrix}.$$

The bifurcation which occurs at $\alpha = 0$ is called the Takens–Bogdanov bifurcation.

A Takens–Bogdanov bifurcation is a codimension-two bifurcation. It can be considered as the coincidence of a saddle–node and a Hopf bifurcation.

*Reference*: Kuznetsov (1995).

**Takens embedding theorem** (*see* **Whitney embedding theorem**)

**tangent bifurcation** (*see* **bifurcation**)

**tangent bundle**
The tangent bundle $TM$ of an $n$-dimensional manifold $M$ is a $2n$-dimensional manifold, the disjoint union of the tangent spaces of $M$. If $U \subset M$ is an open set and $\phi : \mathcal{R}^n \to U$ is a parametrisation of $U$, then

$$\Phi : \mathcal{R}^n \times \mathcal{R}^n \to T_U M,$$

defined by

$$\Phi(x, v) = (\phi(x), D\phi(x)v),$$

is a parametrisation of $T_U M$.

**tangent space**
Let $M$ be a manifold. For each $x \in M$, the set $TM_x$ of all vectors tangent to $M$ at $x$, is called the tangent space.

*Reference*: Arrowsmith and Place (1990).

**Taylor–Couette flow**
Fluid flow between two concentric cylinders which rotate independently is called Taylor–Couette flow. The flow is sensitive to changes in the Reynolds number, produced by adjusting the ratio of their rotation rates, displaying pre-turbulent chaos. The first experiments in this field were carried out by G.I. Taylor in 1923.

*Reference*: Swinney (1983).

**tent map**
The map

$$x_{n+1} = a(1 - x_c)x_n \qquad (x_n \leq x_c)$$

and

$$x_{n+1} = ax_c(1 - x_n) \qquad (x_c \leq x_n \leq 1)$$

where $ax_c(1 - x_c) \leq 1$, is called a tent map (see Figure 121).

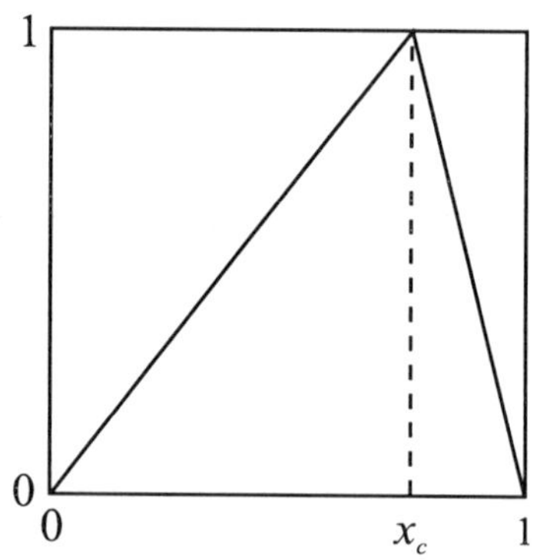


**Figure 121** A tent map

*Remark*: The tent map has a discontinuous derivative at $x = x_c$ so it is not an $S$-unimodal map.

*Example*: Consider the tent map for $a = 4$ and $x_c = 0.5$. Let $x_0 \in [0, 1]$ be the initial point for iterations of the map. In this case one finds

$$x_n = \frac{1}{\pi} \cos^{-1}(\cos 2^n \pi x_0).$$

The iterations show chaotic behaviour with a Lyapunov exponent $\lambda = \ln 2$.

**thermal convection**

Atmospheric flow is caused by the earth's motion and thermal convection. The flow may form patterns such as cells or rolls as hot air rises and cool air falls (*see also* **Benárd cells** and **Lorenz model**).

**Thompson escape equation** (*see* **escape equation**)

**Thom's theorem**

Most smooth infinitely differentiable functions $V(x,c)$, where $x \in \mathcal{R}^n$, $c \in \mathcal{R}^k$, and $k \leq 5$, are structurally stable. For this family $V : \mathcal{R}^n \times \mathcal{R}^k \to \mathcal{R}$ and any point $(x,c) \in \mathcal{R}^n \times \mathcal{R}^k$ there is a choice of coordinates for $c$ in $\mathcal{R}^k$ and for $x \in \mathcal{R}^n$, such that $x$ varies smoothly with $c$, in terms of which the function $V(x,c)$ has one of the following local forms:

a constant plus

$$x_1$$

but not a fixed point

$$x_1^2 + x_2^2 + \ldots + x_r^2 - x_{r+1}^2 - \ldots - x_n^2$$

or non-degenerate fixed point; Morse function

$$x_1^3 + c_1 x_1 + (M)$$

fold catastrophe set

$$\pm(x_1^4 + c_2 x_1^2 + c_1 x_1) + (M)$$

cusp(+), or dual cusp(-)

$$x_1^5 + \sum_{k=1}^{3} c_k x^k + (M)$$

swallowtail,

$$\pm\left(x_1^6 + \sum_{k=1}^{4} c_k x^k\right) + (M)$$

butterfly or dual butterfly

$$x_1^7 + \sum_{k=1}^{5} c_k x^k + (M)$$

wigwam

$$x_1^2 x_2 \pm x_2^3 + c_3 x_1^2 + c_2 x_2 + c_1 x_1 + (N)$$

hyperbolic (+), or elliptic (-) umbilic

$$\pm(x_1^2 x_2 + x_2^4 + c_4 x_2^2 + c_3 x_1^2 + c_2 x_2 + c_1 x_1) + (N)$$

parabolic (and dual) umbilic

$$x_1^2 x_2 \pm x_2^5 + c_5 x_2^3 + c_4 x_2^2 + c_3 x_1^2 + c_2 x_2 + c_1 x_1) + (N)$$

second hyperbolic (+) and second elliptic (-) umbilic

$$\pm(x_1^3 + x_2^4 + c_5 x_1 x_2^2 + c_4 x_2^2 + c_3 x_1 x_2 + c_2 x_2 + c_1 x_1)$$

and symbolic (dual) umbilic.

$(M)$ and $(N)$ are given by

$$(M) = x_2^2 + ... + x_r^2 - x_{r+1}^2 - ... - x_n^2$$

and

$$(N) = x_3^2 + ... + x_r^2 - x_{r+1}^2 - ... - x_n^2.$$

Catastrophe sets of five elementary bifurcations are shown in Figure 122.

*Remark*: Thom's theorem gives eleven elementary catastrophe sets (not counting duals). For $k > 5$, the number of forms is infinite.

**thermohaline convection**

This expression refers to the ocean currents caused by heat and salinity which in turn affect the buoyancy of the fluid (*see also* **El-Niño event**).

**three-body problem**

The aim of the three-body problem, originally posed in celestial mechanics, is to find the general solution of the motion of three bodies all acting under the influence of gravity. This topic provided the stimulation for H. Poincaré's prize-winning essay which was essentially the birth of dynamical systems.

**time-delay method**

This method is a useful technique for evaluating experimental data $x_t$, performed by plotting the time series using the delay coordinates

$$(x(t), x(t-\tau))$$

where $\tau$ is constant (*see also* **delay coordinates reconstruction**).

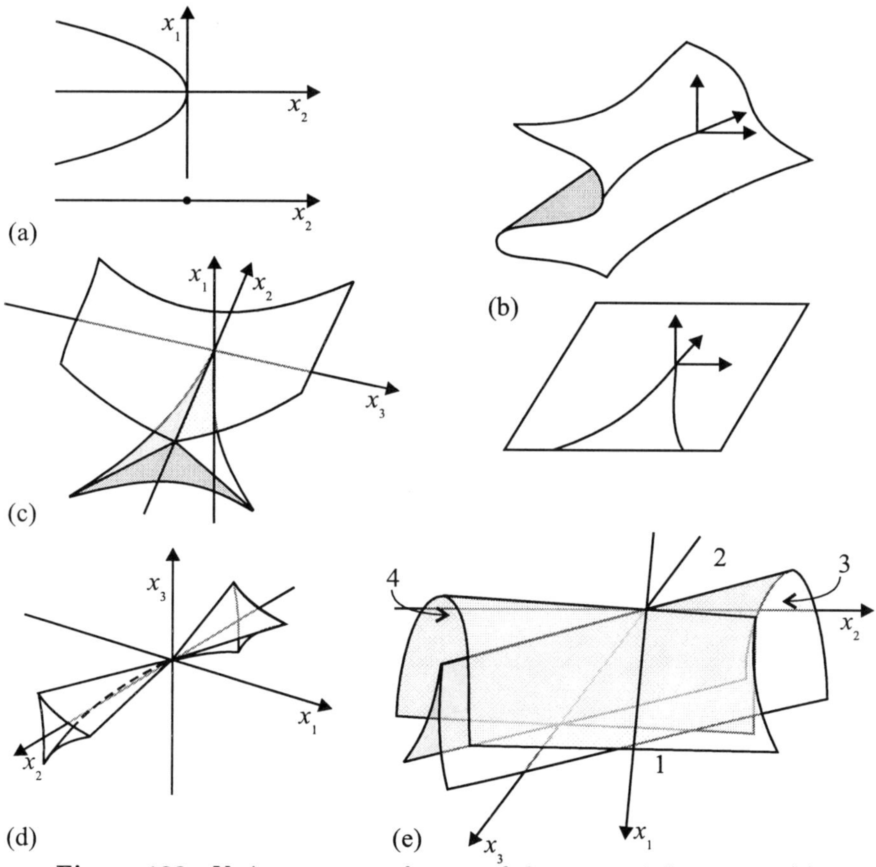


**Figure 122** Various catastrophe sets of elementary bifurcations: (a) fold; (b) cusp; (c) swallowtail; (d) elliptic umbilic; (e) hyberbolic umbillic

**time-series analysis**
The study of dynamical systems based on a time series of measurements is called time-series analysis (*see also* **delay coordinates reconstruction**, **dimension of sets**).

*Reference*: Kantz and Schreiber (1998).

**tinkerbell map**
The map of the plane

$$f(x,y) = (x^2 - y^2 + c_1 x + c_2 y, 2xy + c_3 x + c_4 y)$$

where $c_{1-4}$ are constant, is called the tinkerbell map.

*Reference*: Alligood *et al.* (1997).

**Toda lattice**
This is a model of a monocrystal as a lattice of particles connected by nonlinear springs. The Hamiltonian is given by

$$H(p_i, q_i) = \frac{1}{2}\sum_{n=1}^{n} p_i^2 + \sum_{i=1}^{n-1} \exp(q_i - q_{i+1})$$

where $q_i$ represents the generalised coordinates of the $i$th particle and $p_i$ the canonical momenta.

*Reference*: Jackson (1990).

**topological conjugate**
Two maps, $f : \mathcal{R}^n \to \mathcal{R}^n$, and $g : \mathcal{R}^n \to \mathcal{R}^n$ are said to be topologically conjugate if

$$g = hfh^{-1}$$

for some homeomorphism $h : \mathcal{R}^n \to \mathcal{R}^n$ .

*Example*: Let $g : [0,1] \to [0,1]$ be the logistic map $g(x) = 4x(1-x)$ and let $f : [0,1] \to [0,1]$ be the tent map

$$f(x) = \begin{cases} 2x & : \ 0 \le x \le \frac{1}{2} \\ 2(1-x) & : \ \frac{1}{2} < x \le 1 \end{cases} .$$

It can be shown that the maps $f$ and $g$ are topologically conjugate with

$$h(x) = \frac{2}{\pi} \sin^{-1} \sqrt{x}.$$

**topological dimension** (*see* **dimension of sets**)

**topological entropy**
Let $\Lambda$ be a compact invariant set for a diffeomorphism $f : \mathcal{R}^m \to \mathcal{R}^m$. For an integer $n > 0$ and a number $\epsilon > 0$, a $(n, \epsilon)$ separated set $S \subset \Lambda$ is a set which has the property that $x, y \in S$ and $x \neq y$, implies that there is an integer $0 \le i \le n$ such that the distance

$$d(f^i(x), f^i(y)) > \epsilon.$$

Let $s(n, \epsilon)$ be the maximum cardinality of an $(n, \epsilon)$ separated subset of $\Lambda$. Define

$$h(f, \epsilon) = \lim_{n\to\infty} \sup \frac{1}{n} \ln s(n, \epsilon)$$

and

$$h(f) = \lim_{\epsilon \to 0} h(f, \epsilon).$$

Then $h(f)$ is called the topological entropy of $f$.

Let $\mu$ be an invariant, ergodic, probability measure for $f : \mathcal{R}^n \to \mathcal{R}^n$ with compact support and consider the set

$$V(x, \epsilon, n) = \{y \in \mathcal{R}^n | d(f^i(x), f^i(y)) < \epsilon, 0 \leq i < n\}.$$

Then for almost all $x$ with respect to $\mu$, the number

$$\lim_{\epsilon \to 0} \lim_{n \to \infty} \inf \left\{ -\frac{1}{n} \ln \mu(V(x, \epsilon, n)) \right\} = h_\mu(f)$$

is independent of $x$. The number $h_\mu(f)$ is the $\mu$-entropy of $f$.

*Reference*: Guckenheimer and Holmes (1983).

**topological equivalence**
The set $A$ is said to be topologically equivalent to the set $B$ if there exists homeomorphism $h$ such that

$$h(A) = B$$

i.e. $h$ maps the two sets $A$ and $B$ into each other.

Two $C^r$ maps $F$ and $G$ are $C^k$ equivalent or $C^r$ conjugate ($k \leq r$) if there exists a $C^k$ homeomorphism $h$ such that $hF = Gh$; $C^0$ equivalence is called topological equivalence.

*Reference*: Guckenheimer and Holmes (1983).

**topological invariant**
This is a property which is common to all topologically equivalent sets.

**topological orbital equivalence**
Consider the autonomous system

$$\frac{\mathrm{d}x}{\mathrm{d}t} = f(x)$$

where $x \in \mathcal{R}^n$. Different functions $f(x)$, and corresponding solutions $x^1(t)$, $x^2(t), \ldots$, etc. although analytically quite different, may have a family of trajectories in phase space that possess features which are 'similar' to each other. If one family of trajectories can be continuously deformed into another family, while still retaining the orientation of motion in the phase space, then such systems are called topologically orbitally equivalent.

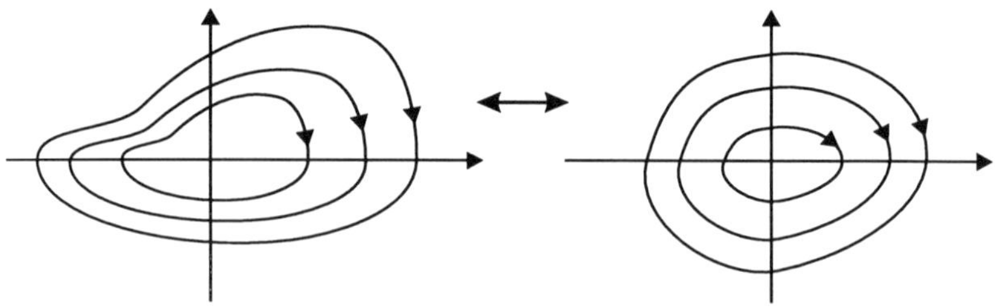

**Figure 123** Illustration of topological orbital equivalence

*Example*: Figure 123 shows two families in $\mathcal{R}^2$ which can clearly be continuously deformed into each other (by stretching or contracting, but not by cutting or connecting orbits).

**torus**

The $n$-dimensional torus is the direct product of $n$ copies of the circle, i.e.

$$T^n = S^1 \times S^1 \times S^1 \times ... \times S^1.$$

*Example*: The two-dimensional torus $T^2 = S^1 \times S^1$, can be represented as the square $\{x, y : 0 \leq x \leq 2\pi, 0 \leq y \leq 2\pi\}$ with opposite sides pasted together, i.e. the points $(0, y)$ and $(2\pi, y)$, as well as the points $(x, 0)$ and $(x, 2\pi)$, are identified.

**trajectory**

The solution $x(t)$ of an ordinary differential equation, $\mathrm{d}x/\mathrm{d}t = f(x)$, where $x \in \mathcal{R}^n$, from a given initial condition $x_0 = x(t = 0)$, plotted in phase space is called a trajectory or orbit (*see* **topological orbital equivalence**).

**transcritical bifurcation** (*see* **bifurcation**)

**transient**

The evolution of a dynamical system before settling to an attractor is termed the transient.

*Example*: A transient evolution converging to the limit cycle attractor is shown in Figure 124.

**transient chaos**

Consider a typical orbit $\gamma$ of a dynamical system in a space $\mathcal{R}^n$ such that as $t \to \infty$ then $\gamma$ approaches a non-chaotic attractor $A$ ($A$ could be $\infty$). If for $t < \tau$ the evolution of the orbit $\gamma$ is on a subset $S$ ($S \cap A = \emptyset$) which has all the hallmarks of chaos, then such an evolutution is called transient chaos or $1/\tau$ transient chaos.

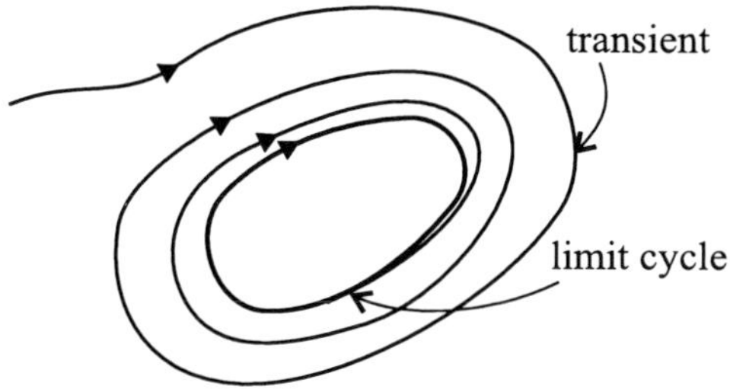


**Figure 124** Transient evolution towards the limit cycle attractor

*Remark*: Transient chaos is characterised by a positive transient maximum Lyapunov exponent.

*Reference*: Alligood *et al.* (1997).

**translation map**
The map $F : \mathcal{R}^2 \to \mathcal{R}^2$, defined by

$$F(x, y) = (x + a, y + b)$$

where $a$ and $b$ are constant, is an area-preserving translation map of the plane.

**transversality**
The transversality theorem of differential topology implies that when two manifolds (surfaces) of dimensions $k$ and $l$ meet in an $n$-dimensional space, then, in general, their intersection will be a manifold of dimension $(n-(k+l))$.

*Remark*: If $k + l < n$, then one does not expect intersections to occur at all.

In general, the meaning of transversality is given locally in terms of tangent spaces.

*Reference*: Guckenheimer and Holmes (1983).

**transverse intersection**
This is an intersection of manifolds such that, from any point in the intersection, all directions in the phase space can be generated by a linear combination of vectors tangent to the manifolds.

*Example*: Consider the intersections of the curves and surfaces in $\mathcal{R}^3$ shown in Figure 125, where bold arrows indicate tangents. We note that in this case their span must be three-dimensional for transversality.

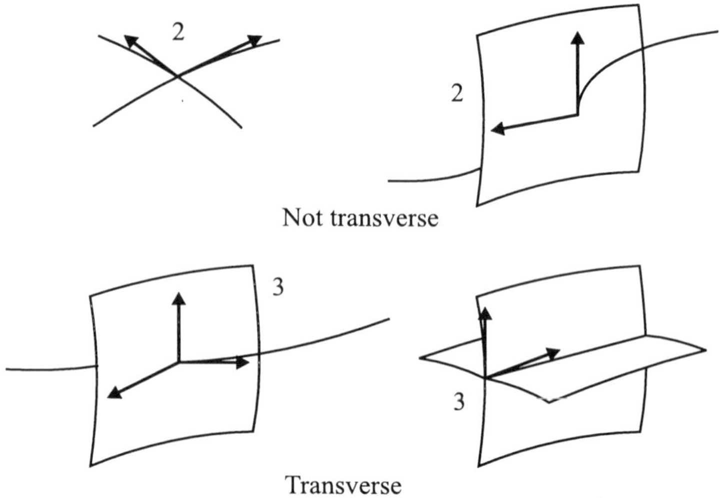


**Figure 125** Transverse intersection of curves and surfaces in $\mathcal{R}^3$

*Reference*: Thompson and Stewart (1986).

**transverse manifold**

Consider a dynamical system

$$\frac{dx}{dt} = f(x)$$

where $x \in \mathcal{R}^n$ with a vector field $f(x)$, near some non-trivial orbit $\gamma$ (not a fixed point). A manifold $\sum$ is called a transverse manifold if the normal to it, $\hat{n}(x)$, is nowhere orthogonal to $f(x)$, for $x \in \sum$, i.e. $f(x)\hat{n}(x) \neq 0$ for all $x \in \sum \subset \mathcal{R}^n$.

**trapping region**

If we consider a flow $\phi_t$ and a set $D$ such that $D \in \mathcal{R}^n$ and $\phi_t(D) \subset D$ for all $t > 0$, then the set $D$ is called the trapping region. Usually, it is sufficient to show that the vector field is directed inwards everywhere on the boundary of $D$.

A similar definition also exists for maps.

*Reference*: Arrowsmith and Place (1992).

**turbulence**

If a physical system consisting of a viscous fluid (and rigid bodies) is not subjected to any external action, it will tend to a state of rest (equilibrium).

We now submit the system to a steady action (pumping, heating, etc.) measured by a parameter $\mu$. When $\mu > 0$, we obtain first a steady state, i.e. the physical parameters describing the fluid at any point (velocity, temperature, etc.) are constant in time. This steady situation prevails for small values of $\mu$. When $\mu$ is increased, various new phenomena occur, as follows:

(i) the fluid motion may remain steady but change its symmetry pattern;

(ii) the fluid motion may become periodic in time;

(iii) for sufficiently large $\mu$, the fluid motion becomes very complicated, irregular and unpredictable i.e. it becomes turbulent.

*Reference*: Ruelle and Takens (1971).

**twist map**

Let $A$ be an annulus

$$A = S^1 \times [a, b] = \{(\phi, r) : 0 \leq \phi \leq 2\pi, a \leq r \leq b\}.$$

A twist map is a homeomorphism $T : A \to A$ which has the form

$$T : (\phi, r) \to (\phi + \alpha(r), r)$$

where $\mathrm{d}\alpha/\mathrm{d}r \neq 0$ for $r \in [a, b]$.

*References*: Arrowsmith and Place (1990); Katok (1982).

# U

**Ueda's equation**
The second-order ordinary differential equation with cubic nonlinearity but no linear term:

$$\frac{\mathrm{d}^2x}{\mathrm{d}t^2} + k\frac{\mathrm{d}x}{\mathrm{d}t} + x^3 = B\cos t$$

where $k$ and $B$ are constants, is called Ueda's equation. It has been much studied by Y. Ueda since the early 1960s.

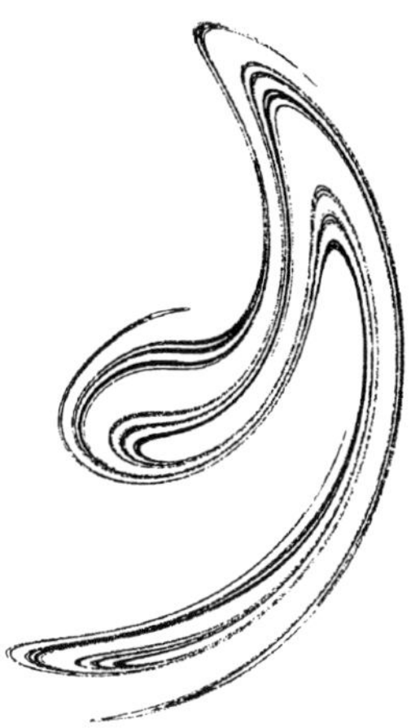

**Figure 126** Japanese attractor

Originally used to model a series-resonance electrical circuit with nonlinear inductance, it was used by Ueda initially to study nonlinear dynamics by using harmonic balance techniques to produce analytic approximations to solutions but these were also compared with analogue computer simulations. Subsequently, these solutions were compared with numerical treatments. The system with $k = 0.5$ and $B = 7.5$ produces a chaotic response whose associated

Poincaré map shown in Figure 126 is sometimes referred to as the Ueda or Japanese attractor.

*Reference*: Ueda (1992).

**ultrasubharmonic**
Consider the forced oscillator

$$\frac{\mathrm{d}^2x}{\mathrm{d}t^2} + f(x, \frac{\mathrm{d}x}{\mathrm{d}t}) = A\cos\Omega t.$$

If the driving frequency $\Omega$ and the oscillation frequency $\omega$ are related in the following way:

$$\omega = \frac{m\Omega}{n}$$

where $m$ and $n$ are integers, and $m \neq 1$, and $n \neq 1$, then the motion is said to be ultrasubharmonic.

**ultrasuperharmonic**
Consider the forced oscillator

$$\frac{\mathrm{d}^2x}{\mathrm{d}t^2} + f(x, \frac{\mathrm{d}x}{\mathrm{d}t}) = A\cos\Omega t.$$

If the driving frequency $\Omega$ and the oscillation frequency $\omega$ are related in the following way:

$$\omega = m\Omega$$

where $m$ is an integer, then the motion is said to be ultrasuperharmonic or super-harmonic.

**umbilic**
One of the elementary catastrophes (*see* **Thom's theorem**).

**unbounded solution** (*see* **escape to infinity**)

**unfolding** (*see* **codimension**)

**unimodal map**
A continuous map

$$f : [-1, 1] \to [-1, 1]$$

such that $f$ is strictly increasing on the interval [-1,0], and strictly decreasing on [0,1], and such that

$$f(-1) = f(1) = -1$$

is called a unimodal map.

A unimodal map is called $S$-unimodal, if it is in the class $C^3$ and if it also has a negative Schwarzian derivative on [-1,1] - { 0 }.

*Example*: Let $f : [-1,1] \to [-1,1]$ be given as

$$f(x) = 1 - 2x^2.$$

Then $f$ is a unimodal map.

*Remark*: The given definition can be generalised to the maps

$$f : [a,b] \to [a,b].$$

In this case, map $f$ is unimodal if there exists $c$, $(a < c < b)$ such that $f$ is increasing in the interval $[a,c]$ and decreasing in the interval $[c,b]$.

**uniquely ergodic**
Let $X$ be a compact metric space. The $\sigma$-algebra of Borel subsets of $X$ will be denoted by $B(X)$. Let $M(X)$ be the collection of all probability measures defined on the measurable space $(X, B(X))$. A map

$$\bar{T} : M(X) \to M(X)$$

may be defined by

$$(\bar{T}\mu)(B) = \mu(T^{-1}B).$$

Now let

$$M(x,T) = \{\mu \in M(X) : \bar{T}\mu = \mu\}.$$

This set consists of all $\mu \in M(X)$, thus making $T$ a measure-preserving transformation of $(X, B(X), \mu)$.

A continuous transformation $T : X \to X$ is called uniquely ergodic if there is only one $T$-invariant Borel probability measure on $X$, i.e. $M(X,T)$ consists of one point.

*Reference*: Walters (1982).

**uniqueness of inversion** (*see* **implicit function theorem**)

**uniqueness of solution** (*see* **existence and uniqueness theorem for solutions of ODEs**)

**unpredictability** (*see* **sensitive dependence on initial conditions**)

**unstable manifold** (*see* **stable and unstable manifolds**)

**U-sequence**
Consider a family of maps

$$x_{n+1} = cf(x_n) = F(x_n, c)$$

where $c$ is a real number such that $0 \leq c \leq c_m$ and $f(x)$ is continuous and differentiable, with a single maximum at $x = x_m$. Consider values of $c$ such that the point $x_m$ is a $p$-periodic point of $F^p(x, c)$, i.e. consider the superstable point $c = \bar{c}_p$ defined by

$$x_m = F^p(x_m, \bar{c}_p).$$

The number of $\bar{c}_p$ values for different values of $p$ is given as

$$1 \text{ (for } p = 2), \ 1(3), \ 2(4), \ 3(5), \ 5(6), \ 9(7), \ 16(8), \ 28(9),$$

for a total of 2370 $\bar{c}_p$ values for $p \leq 15$.

These results were obtained from extensive computer studies carried out by N. Metropolis, M.L. Stein and P.R. Stein. They considered the set of points generated by one of these ($\bar{c}_p$) maps, $x_1 = F(x_m, \bar{c}_p)$, $x_2 = F(x_m, \bar{c}_p)$, ..., etc., starting from $x_m$, and considered whether $x_k > x_m$ (an $R$ point) or $x_k < x_m$ (an $L$ point). Any periodic set is then characterised by some symbolic pattern, such as

$$x_m \to R \to L \to L \to R \to L \to R \to \ldots. \to x_m$$

which contains $(p-1)$ terms $(R, L)$. Patterns of this form were represented by the obvious notation, $RL^2RLR$... . It can be shown that the sequence of these patterns and hence the ordering of periods $p$, as $c$ is increased though the values contained in the set $\{\bar{c}_p\}$, is independent of the function $f(x)$, for a large class of functions. This sequence is called the $U$-sequence.

*Example*: For $p \leq 7$, we have the following sequences

$$R, \ RLR, \ RLR^3, \ RLR^4, \ RLR^2, \ RLR^2LR, \ RL, \ RL^2RL,$$
$$RL^2RLR, \ RL^2R, \ \ldots \ (+ \ 11 \text{ more})$$

where

$$p = (1 + \text{sum of exponents}) \leq 7.$$

*Reference*: Metropolis *et al.* (1973).

# V

**van der Pol equation**
The equation

$$\frac{\mathrm{d}^2 y}{\mathrm{d}t^2} + \delta(1 - y^2)\frac{\mathrm{d}y}{\mathrm{d}t} + \omega^2 y = 0 \qquad (1)$$

where $\delta$ and $\omega$ are constants, is called the van der Pol equation.

The forced system

$$\frac{\mathrm{d}^2 y}{\mathrm{d}t^2} + \delta(1 - y^2)\frac{\mathrm{d}y}{\mathrm{d}t} + \omega^2 y = \rho\cos(\Omega t) \qquad (2)$$

where $\rho$ and $\Omega$ are respectively the amplitude and frequency of the excitation force, is called the forced van der Pol equation and was first developed by this worker to model the behaviour of simple electrical resistance–capacitance (RC) circuits.

*Remark*: Chaotic behaviour can be presented only as the forced van der Pol equation, as according to the Poincaré–Bendixon theorem the only possible $\omega$-limit sets of the autonomous van der Pol equation are a fixed point or a limit cycle.

*Example*: For $\delta = 5$, $\rho = 5$, $\omega = 2.466$ and $\Omega = 1$, the forced van der Pol equation (2) evolves on the chaotic attractor shown in Figure 127.

**variational equation** (*see* **linearisation**)

**variational principle for entropy**
If $T$ is a continuous map of a compact set $S$, then the topological entropy $h_{top}(T) = \sup h_\mu(T)$, where $\sup h_\mu(T)$ represents the measure theoretic entropy with respect to the invariant measure $\mu$.

*Reference*: Walters (1982).

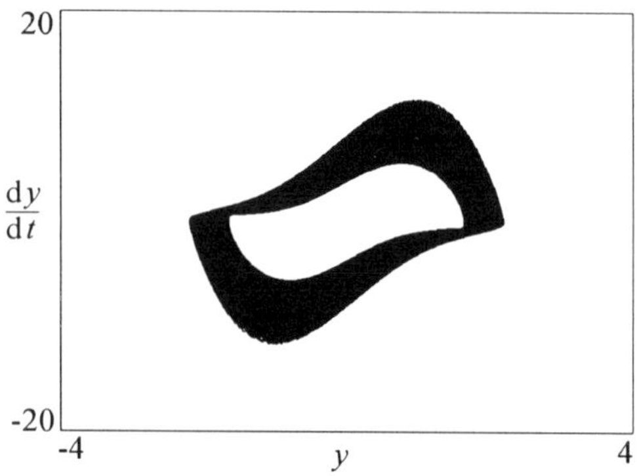


**Figure 127** Chaotic attractor of the forced van der Pol equation

**vector field**
Let $M$ be a $C^{\infty}$ manifold, and for each point $p \in M$, let $X_p$ be a tangent vector to $M$ at $p$. The correspondence $X : p \to X_p$ is called a vector field.

*Example*: Consider the Rössler equations (*see* **Rössler system**). In this case, $M = \mathcal{R}^3$ and the vector field $V$ is given by

$$V = (-y - z)\frac{\partial}{\partial x} + (x + ay)\frac{\partial}{\partial y} + (b + z(x - c))\frac{\partial}{\partial z}.$$

**vibro-impact system**
This is the term used to represent a vibrating system which also undergoes repeated impacts, being historically studied predominantly in the former Soviet Union. Renewed interest in such systems concentrate on the behaviour when impacting first begins at the so-called grazing bifurcation (*see also* **impact oscillator**).

**viscoelastic structures**
In the study of nonlinear vibrations of continuum systems, such as beams, strings and plates, the response of a deformable body for small oscillations can be adequately described by linear equations together with appropriate boundary conditions. For larger-amplitude motions, nonlinearities will come into play which can be geometrical, inertial or material in nature.

**Volterra–Lotka equations** (*see* **Lotka–Volterra equations**)

**volume-preserving map**
Consider a set $S_n \subset \mathcal{R}^m$ and a continuous differentiable map $F : \mathcal{R}^m \to \mathcal{R}^m$; we may then define $S_{n+1}$ as the set $F(S_n)$, i.e. the points $x_{n+1} = F(x_n)$ for

all $x_n \subset S_n$. In addition, define $\mu_n$ as the Lebesgue measure (hypervolume) of the set $S_n$. The local ratio of $\mu_{n+1}$ to $\mu_n$ is given by

$$\frac{\mu_{n+1}}{\mu_n} = |\det J(x_n)|$$

where $J$ is the $m \times m$ Jacobian matrix whose elements are $\partial F_i / \partial x_j$.

If the modulus $|\det J| < 1$ for all $x \in \mathcal{R}^m$, then the hypervolume $\mu$ decreases monotonically as $n$ increases, and the volume shrinks to zero, thus producing an attractor whose dimension is less than $m$.

If the modulus $|\det J| = 1$ for all $x \in \mathcal{R}^m$, then $\mu_n = \mu_0$ for all $n$, i.e. the hypervolume $\mu_n$ is independent of $n$ and we say that $F$ is a volume-preserving map. More precisely, if $m = 2$ we say area-preserving, and if $m = 3$ volume-preserving; otherwise, the more general term measure-preserving can be applied.

*Reference*: Drazin (1992).

**vortex**
This term is sometimes used as another name for centre (*see* **fixed points**).

# W

$\omega$-**limit set** (*see* **omega limit set**)

**Wada property**
This is a term used by C. Grebogi, J.A. Yorke and their co-workers from the University of Maryland. A basin of attraction $B$ in phase space is said to have the Wada property if there exists two other basins $B_2$ and $B_3$ such that every point on the boundary of $B$ is also on the boundary of $B_2$ and $B_3$.

*Reference*: Alligood *et al.* (1997).

**wandering point**
A point $x$ wanders under a mapping $f$ if it has a neighbourhood $U$, such that $f^n(U) \cap U \neq 0$ for all $n \geq 0$.

**waterwheel**
A model of a waterwheel (also called the Lorenzian waterwheel) describes thermally induced fluid convection in the atmosphere. Fluid heated from below becomes lighter and rises, whereas heavier fluid falls under gravity. Such motions often produce convection rolls similar to the motion of the fluid in a circular torus. The waterwheel was one of the first examples of chaos in a mechanical device (*see also* **Rayleigh–Bénard convection**).

*References*: Lorenz (1993); Moon (1992).

**wave equation**
The partial differential equation used in the study of waves (acoustic, fluid, electromagnetic, etc.) given by

$$\frac{\partial^2 y}{\partial t^2} = c^2 \frac{\partial^2 y}{\partial x^2}$$

where $x \in \mathcal{R}^n$ and $c$ is a constant, is called the wave equation.

**Weierstrass function**
The function

$$f(x) = \sum_{i=1}^{\infty} \lambda^{(s-2)i} \sin(\lambda^i x)$$

where $1 < s < 2$ and $\lambda > 1$, is called the Weierstrass function.

*Remark*: The Weierstrass function is continuous but nowhere differentiable.

Weierstrass functions are used to describe fractals.

**Whitney embedding theorem**
This theorem relates to the problem of reconstructing state space from a time series of measurements. Assume that the state space is $\mathcal{R}^k$ and that trajectories are attracted towards a $d$-dimensional manifold $M$. For each state we assume that we can make $m$ simultaneous independent measurements at any given time.

If we associate a function $F$ with the measuring process at different times then $F : \mathcal{R}^k \to \mathcal{R}^m$. We can evaluate $F$ at any point of $M \in \mathcal{R}^K$ by carrying out $m$ measurements and then producing a vector from them.

If $m > 2d$, and $F : \mathcal{R}^k \to \mathcal{R}^m$ is generic, then $F$ is a one-to-one map on $A$.

Let $M$ be a smooth compact manifold of dimension $d$. It is then a generic property of smooth maps $G : M \to \mathcal{R}^{2d+1}$ that $G$ is an embedding.

This result is valid for generic maps from $M \to \mathcal{R}^{(2d+1)}$ but says nothing about specific maps or even any specific subsets of maps. F. Takens also considered the restriction of $G$ to be delay maps reconstructed from a smooth observation function $h$ and the dynamics $f$, i.e.

$$f = \{h(x), h(f(x)), ...h(f^{2d}(x))\}.$$

Although generally these results are used to justify the use of embedding for general maps, care should be taken when working with a specific set of maps that do not produce embeddings (e.g. symmetry).

*References*: Whitney (1936); Takens (1981).

**wigwam**
One of the elementary catastrophes (*see* **Thom's theorem**).

**winding number**
Consider a circle map

$$\theta_{n+1} = [\theta_n + \Omega - \frac{K}{2\pi} \sin(2\pi\theta_n)]$$

where $K, and\Omega$ are constants and the square bracket indicates that mod $2\pi$ of the enclosed expression is taken.

The winding number $W$ for an orbit $\{\theta_n\}$ is given by

$$W(K,\Omega) = \lim_{n\to\infty} \frac{\theta_n - \theta_0}{n}.$$

*Remark*: The winding number $W(K,\Omega)$ is related to the rotation number $\rho(K,\phi)$ in the following way. Let $\phi_n = 2\pi\theta_n$ and $\Phi = 2\pi\Omega$, then

$$W(K,\Omega) = \rho(K,\Phi).$$

The rotation number $2\pi\rho$ represents the time-average angular rate of rotation of the orbit $\{\phi_n\}$ circles around in the $\phi$ direction.

*Example*: Consider a map

$$\zeta_{n+1} = [\zeta_n + 2\pi\omega] \bmod 2\pi.$$

The winding number $W$ for an orbit $\{\zeta_n\}$ is equal to $\omega$, since

$$\zeta_n = \zeta_0 + 2\pi n\omega.$$

# Z

**Zeeman's catastrophe machine**
This machine is a simple mechanism which is used as an example in catastrophe theory. It consists of a disc of radius $r$, pivoted at its centre, at which friction acts. Two rubber bands, each with an unstretched length $l_0$, are attached to a perpendicular peg at the edge of the disc. The other end of one of the bands is attached to the point $(-d, 0)$ in the plane of the disc $(x, y)$, where $d > l_0 + r$. This ensures that if the angle of the peg from the horizontal $\theta$ is such that if $\theta = \pi$, then this rubber band is under tension. The end of the other rubber band is put at some arbitrary control point in this plane. There are two control parameters, $(r, \theta)$ or $(r_c, \theta_c)$, as shown in Figure 128. Experiments show that as $r_c$ is decreased, and $\theta_c$ is moved from positive to negative values, and back again, the value of $\theta$ exhibits a jump phenomenon, as illustrated in Figure 129. The stretched lengths of the rubber bands are $l$ and $l_c$, respectively.

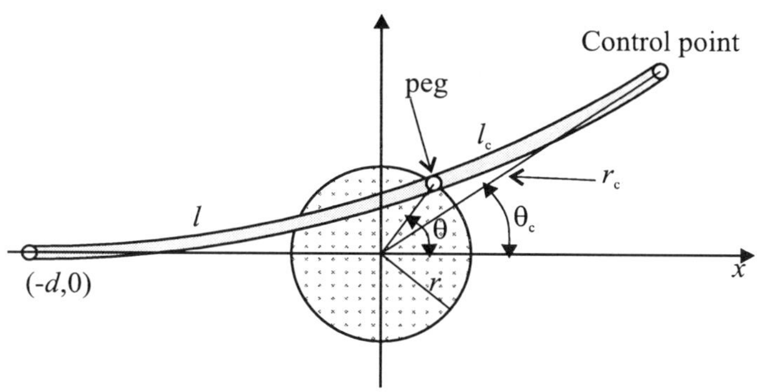


**Figure 128** Schematic representation of Zeeman's catastrophe machine

The values of $(r_c, \theta_c)$ where jumps occur are on two different curves, depending on whether $\theta_c$ is increasing or decreasing. Moreover, these curves intersect in

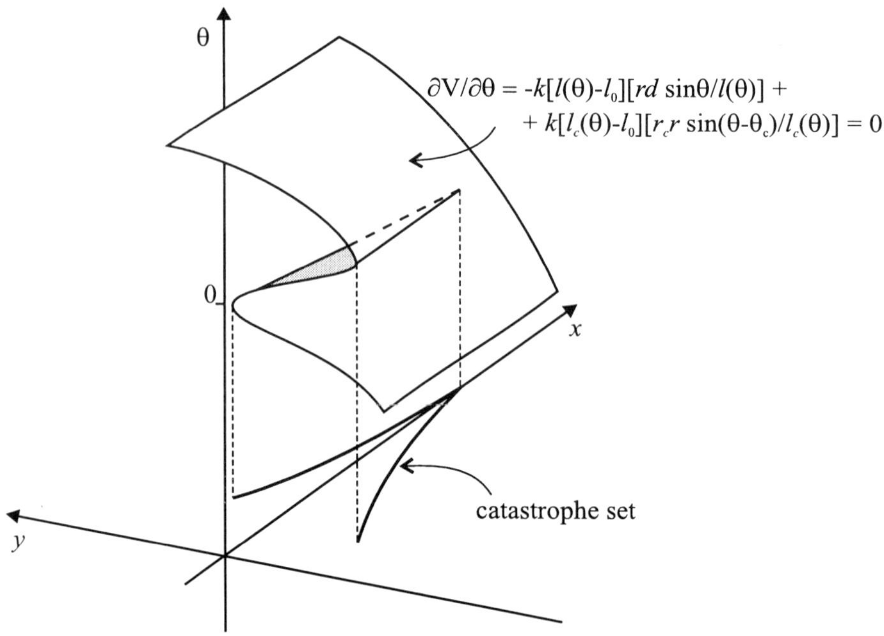


**Figure 129** Illustration of the cusp catastrophe in Zeeman's machine

a cusp, and so locally the equilibrium surface is a cusp-catastrophe surface.

*References*: Arnold (1984), Jackson (1990), Zeeman (1977).

# References


Abraham, R.H. and Shaw, C.D. (1984) *Dynamics: The Geometry of Behaviour*, Part III, Ariel Press, Santa Cruz.

Alexander, J.C., Kan, I., Yorke, J.A. and You, Z. (1992) 'Riddled basins', *Int. J. Bif. Chaos.*, **2**, 795–812.

Alligood, K.T., Sauer, T.D. and Yorke, J.A. (1997) *Chaos: An Introduction to Dynamical Systems*, Springer, New York.

Andronov, A.A., Vitt, E.A. and Khaiken, S.E. (1966) *Theory of Oscillations*, Pergamon Press, Oxford.

Arnold, L. (1974) *Stochastic Differential Equations: Theory and Applications*, Wiley, New York.

Arnold, V.I. (1980) *Mathematical Methods of Classical Mechanics*, Springer, New York.

Arnold, V.I. (1983) *Geometrical Methods in the Theory of Differential Equations*, Springer, New York.

Arnold, V.I. (1984) *Catastrophe Theory*, Springer, New York.

Arnold, V.I. (Ed.) (1988) *Dynamical Systems III*, Springer, Berlin.

Arnold, V.I. and Avez, Z. (1968) *Ergodic Problems of Classical Mechanics*, Benjamin, New York.

Arrowsmith, D.K. and Place, C.M. (1990) *An Introduction to Dynamical Systems*, Cambridge University Press, Cambridge.

Arrowsmith, D.K. and Place, C.M. (1992) *Dynamical Systems*, Chapman and Hall, London.

Ashwin, P., Buescu, J. and Stewart, I. (1994) 'Bubbling of attractors and synchronization of chaotic attractors', *Phys. Lett.*, **A193**, 126–134.

Bak, P., Tang, C. and Wiesenfeld, K. (1987) 'Self-organized criticality: An explanation of 1/f noise', *Phys. Rev. Lett.*, **59**, 381–384.

Barnsley, M. (1988) *Fractals Everywhere*, Academic Press, Boston.

Becker, K.-H. and Dorfler, M. (1989) *Dynamical Systems and Fractals*, Cambridge University Press, Cambridge.

Benettin, G., Galgani, L., Giorgilli, A. and Strelcyn, J.-M. (1980) 'Lyapunov characteristic exponents for smooth dynamical systems and for Hamiltonian systems: A method of calculating all of them', *Meccanica*, **15**, 9–27.

Benzi, R., Parisi, G., Sutera, A. and Vulpiani, A. (1982) 'Stochastic resonance in climatic change', *Tellus*, **34**, 10–16.

Berry, M. (1989) 'Quantum chaology', *Physica Scripta*, **40**, 335–346.

Billingsley, P. (1965) *Ergodic Theory and Information*, John Wiley & Sons, New York.

Busse, F.H. (1980) 'Convection in rotating layer', *Science*, **208**, 173–175.

Byrd, P.F. and Friedman, M.D. (1971) *Handbook of Elliptic Integrals for Engineers and Scientists*, Springer, New York.

Carr, J. (1981) *Applications of Center Manifold Theory*, Springer, New York.

Chaitin, G.J. (1987) *Information, Randomness and Incompleteness*, World Scientific, Singapore.

Chow, S.N. and Hale, J. (1982) *Methods of Bifurcation Theory*, Springer, New York.

Chua, L.O. (1992) 'The genesis of Chua's circuit', *Archiv. Elektronik Ubertragungstechnik*, **46**, 250–257.

Chua, L.O., Kocarev, L., Eckert, K. and Itoh, M. (1992) 'Experimental chaos synchronization in Chua's circuit', *Int. J. Bif. Chaos*, **2**, 705–712.

Collet, P., Eckmann, J. P. (1980) *Iterated Maps of the Interval as Dynamical Systems*, Birkhauser, Boston.

Collet, P., Eckmann, J.P. and Lanford, O.E. (1970) 'Universal properties of maps on an interval', *Comm. Math. Phys.*, **76**, 211–254.

Cuomo, K. and Oppenheim, A. (1993) 'Circuit implementation of synchronized chaos with applications to communication', *Phys. Rev. Lett.*, **71**, 65–69.

Dankowicz, H. (1997) *Chaotic Dynamics in Hamiltonian Systems with Applications to Celestial Mechanics*, World Scientific, Singapore.

Devaney, R.L. (1989) *An introduction to Chaotic Dynamical Systems*, W.H.Benjamin, New York.

Ditto, W.L., Rauseo, S.W. and Spano, M.L. (1991) 'Experimental control of chaos', *Phys. Rev. Lett.*, **65**, 3211–3216.

Drazin, P.G. and Johnson, R.S. (1989) *Solitons: An Introduction*, Cambridge University Press, Cambridge.

Drazin,P. (1992) *Nonlinear Systems*, Cambridge University Press, Cambridge.

Dressler, V. and Nitsche, G. (1992) 'Controlling chaos using time delay coordinates', *Phys. Rev. Lett.*, **68**, 1–5.

Ebeling, W. and Peschel, M. (Eds) (1985) *Approach to Cooperation and Competition in Dynamical Systems*, Akademia-Verlag, Berlin.

Eckmann J.-P. and Ruelle, D. (1985) 'Ergodic theory of strange attractors', *Rev. Mod. Phys.*, **57**, 617–659.

Edgar, G.A. (1990) *Measure, Topology and Fractal Geometry*, Springer, New York.

El Naschie, M.S. and Al Athel, S. (1989) 'On the connection between statical and dynamical chaos', Z. Naturforsch., **44a**, 645–652.

El Naschie, M.S. (1989) 'Generalized bifurcation and shellbuckling as spacial statical chaos', it ZAMM, **69**, T367–377.

Falconer, K.J. (1985) *The Geometry of Fractal Ssets*, Cambridge University Press, Cambridge.

Feigenbaum, M. (1978) 'Qualitative universality for a chaos of nonlinear transformations', *J. Stat. Phys.*, **19**, 5–32.

Fermi, E. *Collected papers, vol.II*, University of Chicaga Press, 1965, Chicago.

Foale, S. and Bishop, S.R. (1994) 'Bifurcations in impact oscillators', *Nonlinear Dyn.*, **6**, 285–299.

Gardiner, C.W. (1990) *Handbook of Stochastic Methods*, Springer, Berlin.

Ghrist, R.W. and Holmes, P. (1996) 'An ODE whose solutions contain all knots and links', *Int. J. Bif. Chaos*, **6**, 779–800.

Glass, L. and M.C. Mackey, (1988) *From Clocks to Chaos*, Princeton University Press, Princeton, NJ.

Grassberger, P. and Procacia, J. (1983) 'Measuring the strangeness of strange attractors', *Physica*, **9D**, 189–204.

Gray, P. and Scott, S. (1986) 'Oscillations in forced chemical reactions', *Ber. Bunsenges. Phys. Chem.*, **90**, 985–997.

Grebogi, C., Ott, E. and Yorke, J. A. (1982) 'Chaotic attractors in crisis', *Phys. Rev. Lett.*, **48**, 1507–1510.

Grebogi, C. and Yorke, J.A., (1997) *The Impact of Chaos on Science and Society*, United Nations University Press, Tokyo.

Greenspan, B.D. and Holmes, P. (1984) 'Repeated resonance and homoclinic bifurcations in a periodically forced family of oscillators', *SIAM J. Math. Anal.*, **15**, 69–97.

Guckenheimer, J. and Holmes, P. (1983) *Nonlinear Oscillations, Dynamical Systems and Bifurcations of Vector Fields*, Springer, New York.

Hagedorn, P. (1981) *Nonlinear Oscillations*, Oxford University Press, Oxford.

Haken, H. (1976) *Synergetics*, Springer, Berlin.

Hale, J. (1969) *Ordinary Differential Equations*, John Wiley & Sons, New York.

Hayashi, C. (1964) *Nonlinear Oscillations in Physical Systems*, McGraw-Hill, New York.

Hausdorff, F. (1919) 'Dimension und ausseres mass', *Mathematische Annalen*, **79**, 157–179.

Hénon, M. (1976) 'Two dimensional mapping with strange attractor', *Commun. Math. Phys.*, **50**, 69–77.

Hénon, M. and Heiles, C. (1964) 'The applicability of the third integral of motion: some numerical experiments', *Astron. J.*, **69**, 73–79.

Hirsch, M. and Smale, S. (1974) *Differential Equations, Dynamical Systems and Linear Algebra*, Academic Press, London.

Holmes, C. and Holmes, P. (1981) 'Second order averaging and bifurcations to subharmonics in Duffing's equation', *J. Sound Vib.*, **78**, 161–174.

Holmes, P. (1982) 'The dynamics of repeated impacts with sinusoidally vibrating table', *J. Sound Vib.*, **84**, 173–189.

Hunt, E.H. (1991) 'Stabilizing high-periodic orbits in a chaotic system: the diode resonator', *Phys. Rev. Lett.*, **67**, 1953–1959.

Huntley, I.D. and Johnson, R.M. (1983) *Linear and Nonlinear Differential Equations*, Ellis Horwood, Chichester.

Hsu, C.S. (1987) *Cell-to-Cell Mapping*, Springer, New York.

Irwin, M.C. (1980) *Smooth Dynamical Systems*, Academic Press, London.

Iyanaga, S. and Kawada, Y. (Eds) (1980) *Encyclopedic Dictionary of Mathematics*, MIT Press, Cambridge.

Jackson, E.A. (1990) *Perspectives of Nonlinear Dynamics, Vols 1 and 2*, Cambridge University Press, Cambridge.

Jordan, D.W. and Smith, P. (1987) *Nonlinear Ordinary Differential Equations*, Oxford University Press, Oxford.

Kaneko, K. (1993) *Theory and Applications of Coupled Map Lattices*, John Wiley & Sons, Chichester.

Kantz, H. and Schreiber, T. (1998) *Nonlinear Time Series Analysis*, Cambridge University Press, Cambridge.

Kapitaniak, T. (1990) *Chaos in systems with noise*, World Scientific, Singapore.

Kapitaniak, T. (1991) *Chaotic Oscillations in Mechanical Systems*, Manchester University Press, Manchester.

Kapitaniak, T. (1996) *Controlling Chaos*, Academic Press, London.

Kapitaniak, T. and Wojewoda, J. (1993) *Attractors of Quasiperiodically Forced Systems*, World Scientific, Singapore.

Kapitaniak T., Chua, L.O., Zhong, G.Q. (1994) 'Experimental synchronization of chaos using continuous control', *Int. J. Bif. Chaos*, **4**, 483–488.

Kaplan, W. (1973) *Advanced calculus*, Addison-Wesley, Reading, MA.

Kaplan, J.L., Yorke, J.A. (1979) 'Chaotic behaviour of multidimensional difference equations', in: *Functional Differential Equations and Approximations of Fixed Points*, H.-O. Peitgen and H.-O. Walter (Eds), Lecture Notes in Mathematics, **730**, Springer, Berlin.

Katok, A.B. (1982) 'Some remarks on Birkhoff and Mather twist map theorems', *Ergod. Theory Dyn. Syst.*, **2**, 185–194.

Kloeden, P. and Platen, E. (1992) *Numerical Solutions of Stochastic Differential Equations*, Springer Verlag, New York.

Kobrinskii, A.E. and Kobrinskii, A.A. (1973) *Vibroshock systems (Dynamics and Stability)*, Nauka, Moscow (in Russian).

Kocarev, L. Halle, K., Eckert, K. and Chua, L.O. (1992) 'Experimental demonstration of secure communications via chaos synchronization', *Int. J. Bif. Chaos*, **2**, 709–716.

Kolmogorov, A.N. (1958) 'A new metric invariant of transitive dynamical systems', *Dokl. Akad. Nauk SSSR*, **119**, 861–918.

Kunick, A., Steeb, W.–H. (1987) 'Chaos in systems with limit cycles', *Int. J. Nonlinear Mech.*, **22**, 349–362.

Kuznetsov, Yu.A., (1995) *Elements of Applied Bifurcation Theory*, Springer, New York.

Landau, L.D. (1944) 'On the Problem of Turbulence', *C. R. Acad. Sci. URSS*, **44**, 311–318.

Lasota, A. and Mackey, M.C. (1985) *Probabilistic Properties of Deterministic Systems*, Cambridge University Press, Cambridge.

Lasota, A. and Mackey, M. (1994) *Chaos, Fractals, Noise*, 2nd edition, Springer, New York.

Lasker, J. (1989) 'A numerical experiment on the chaotic nature of the Solar system', *Nature*, **38**, 237–238.

Li, T.Y. and Yorke, J.A. (1975) 'Period three implies chaos', *Am. Math. Monthly*, **82**, 985–990.

Lorenz, E. (1963) 'Deterministic nonperiodic flow', *J. Atmos. Sci.*, **2**, 130–141.

Lorenz, E., (1993) *The Essence of Chaos*, UCL Press, London.

Madan, R. (1993) *Chua's Circuit: A Paradigm for Cchaos*, World Scientific, Singapore.

Mane, (1987) *Ergodic Theory and Differentiable Dynamical Systems*, Springer.

May, R. (1976) 'Simple Mathematical models with very complicated dynamics', *Nature*, **26**, 459–467.

McCauley, J.L. and Palmore, J.I. (1986) *Computable Chaotic Orbits*, *Phys. Lett.*, **A15**, 433–439.

McDonald, S.W., Grebogi, C., Ott, E. and Yorke, J.A. (1985) 'Fractal basin boundaries', *Physica*, **17D**, 125–149.

Mandelbrot, B. (1982) *The Fractal Geometry of Nature*, Freeman, San Francisco.

Matsumoto, T. (1984) 'A chaotic attractor from Chua's circuit', *IEEE Trans. Circuits Syst.*, **CAS-31**, 1055–1058.

Mehta, N., Henderson, R. (1991) 'Controlling chaos to generate a periodic orbit', *Phys. Rev., 1991*, **44A**, 4861–4869.

Melnikov, V.K. (1963) 'On the stability of the center for time periodic perturbations', *Trans. Moscow Math. Soc.*, **12**, 1–57.

Merkin, J., Needham, D. and Scott, S. (1986) 'Oscillatory chemical reactions in closed vessels', *Proc. R. Soc. Lond*, **A406**, 299–318.

Metropolis, N., Stein, M.L. and Stein, P.R. (1973) 'On finite sets for transformations on the unit interval', *J. Combin. Theor.*, **A15**, 25–44.

Milnor, J. (1985) 'On the concept of attractor', *Commun. Math. Phys.*, **99**, 177–195.

Milnor, J. and Thurston, R. (1977) *On Iterated Maps of the Interval I and II*, Unpublished notes, Princeton University Press, Princeton.

Moon, F.C., (1992) *Chaotic and Fractal Dynamics*, John Wiley & Sons, Chichester.

Moon, F.C., (1987) *Chaotic Vibrations*, John Wiley & Sons, Chichester.

Morosov, A.D. (1973) 'Approach to a complete qualitative study of Duffing's equation', *USSR J. Comp. Math. and Math. Phys.*, **13**, 1134–1152.

Morton, K.W. and Mayers, D.F. (1994) *Numerical Solution of Partial Differential Equations*, Cambridge University Press, Cambridge.

Moser, J. (1973) *Stable and Random motions in Dynamical Systems with Special Emphasis on Celestial Mechanics*, Princeton University Press, Princeton.

Mullin, T. (1993) *Nature of Chaos*, Oxford University Press, Oxford.

Nayfeh, A.H. and Mook, D.T., (1979) *Nonlinear Oscillations*, John Wiley & Sons, Chichester.

Nayfeh, A.H. and Balachandran, B., (1995) *Applied Nonlinear Dynamics*, John Wiley & Sons, Chichester.

Nayfeh, A.H. (1985) *Problems in Perturbation Theory*, John Wiley & Sons, Chichester.

Newhouse, S., Ruelle, D. and Takens, F. (1978) 'Occurence of strange axiom-A attractors near quasiperiodic flows on $T^m$, $m > 3$', *Commun. Math. Phys.*, **64**, 35–40.

Newhouse, S., (1980) 'Lectures on dynamical systems' in: (Ed(s)) *Dynamical Systems, CIME Lectures Bressanone*, Progress in Mathematics, **8**, Birkhauser, Boston.

Nitecki, Z. (1971) *Differential Dynamics*, MIT Press, Boston.

Nordmark, A.B. (1991) 'Non-periodic motion caused by grazing incidence in an impact oscillator', *J. Sound Vib.*, **145**, 275–297.

Oseledec, V.I. (1968) 'A multiplicative ergodic theorem', *Trans. Moscow Math. Soc.*, **19**, 197–234.

Ott, E. (1992) *Chaos in Dynamical Systems*, Cambridge University Press, Cambridge.

Ott, E., Grebogi, C. and Yorke, J.A. (1990) 'Controlling chaos', *Phys. Rev. Lett.*, **64**, 1196–1199.

Palis, J. and de Melo, W. (1982) *Geometric Theory of Dynamical Systems*, Springer, New York.

Parker, T. and Chua, L.O. (1989) *Practical Numerical Algorithms for Chaotic Systems*, Springer, New York.

Parlitz, V. and Lauterborn, W. (1987) 'Period-doubling cascades and devil's stairase of the driven van der Pol oscillator', *Phys. Rev.*, **36A**, 1428–1434.

Pecora, L. and Carroll, T. (1990) 'Synchronization in chaotic systems', *Phys. Rev. Lett.*, **64**, 821–825.

Peitgen, H.-P., Jürgens, H. and Saupe, D. (1992) *Chaos and Fractals: New Frontiers of Science*, Springer, New York.

Peixoto, M.M. (Ed.) *Dynamical Systems*, Academic Press, London.

Plykin, R., (1974) 'Sources and sinks for A-diffeomorphisms', *USSR Math. Sb.*, **23**, 233–253.

Pomeau, Y., Manneville, P. (1980) 'Intermittent transition to turbulence in dissipative dynamical systems', *Commun Math. Phys.*, **74**, 189–197.

Popp, K. and Stelter, P. (1990) 'Stick-slip vibrations and chaos, *Phil. Trans. R. Soc. London*, **A332**, 89–106.

Poston, T. and Stewart, I. (1978) *Catastrophe Theory and its Applications*, Pitman, London.

Press, W.H., Flannery, B.F., Teukolsky, S.A. and Vetterling, W.T. (1986) *Numerical Recipes*, Cambridge University Press, Cambridge.

Prigogine, I. and Lefever, R. (1968) 'Symmetry breaking instabilities in dissipative systems', *II. J. Chem. Phys.*, **48**, 1695–1700.

Pyragas, K. (1992) 'Continuous control of chaos by self-controlling feedback', *Phys. Lett.*, **170A**, 421–427.

Qu, Z., Hu, G. and Ma, B. (1993) 'Note on continuous chaos control', *Phys., Lett.*, **178A**, 265–272.

Rapp, P.E. (1986) 'Oscillations and chaos in cellular metobolism and physiological systems', in: *Chaos*, A.V. Holden (Ed.), Princeton University Press, NJ.

Rasband, S.N. (1990) *Chaotic Dynamics of Nonlinear Systems*, John Wiley & Sons, New York.

Rietman, E. (1989) *Exploring the Geometry of Nature*, Windcrest, Blue Ridge.

Rollins, R. W. and Hunt, E. R. (1982) 'Intermittent triansiet chaos at interior crises in the diode resonator', *Phys. Rev.*, **29A**, 3327–3334.

Romeiras, F. and Ott, E. (1987) 'Strange nonchaotic attractors of the damped pendulum with quasiperiodic force', *Phys. Rev.*, **A35**, 4404–4412.

Romeiras, F., Ott, E., Grebogi, C. and Dayawansa, W.P. (1992) 'Controlling chaotic dynamical systems', *Physica*, **58D**, 165–180.

Rössler, O.E. (1976) 'An equation for continuous chaos', *Phys. Lett.*, **57A**, 397–399.

Rössler, O.E. (1983) 'The chaotic hierarchy', *Z. Naturforsch.*, **38a**, 788–801.

Ruelle, D. (1981) 'Small random perturbations of dynamical systems and the definition of attractors', *Comm. Math. Phys.*, **82**, 137–151.

Ruelle, D. and Takens, F. (1971) 'On the nature of turbulence', *Comm. Math. Phys.* **20**, 167–192; **23**, 343–344.

Sato, M., Sano, M. and Sawada, Y. (1983) 'Universal scaling property in bifurcation structure of Duffing's and generalized Duffing's equations', *Phys. Rev.*, **82A**, 1654– 1658.

Schroeder, M.R. (1984) *Number Theory in Science and Communication*, Springer, Berlin.

Schuster, H.G. (1984) *Deterministic Chaos*, Physik-Verlag, Weinheim.

Sharkovsky, A.N. (1964) 'Coexistence of cycles of a continuous map of a line into itself', *Ukr. Math. Z.*, **16**, 61–71.

Sel'kov, E.E. (1968) 'Self-oscillations in glycolysis, Part. 1. A simple kinetic models', *European J. Biochem.*, **4**, 79–98.

Shaw, S.W. and Holmes, P.J. (1982) 'A periodically forced linear oscillator', *J. Sound Vib.*, **90**, 129–155.

Singer, J., Wang, Y.-Z. and Bau, H.H. (1991) 'Controlling a chaotic system', *Phys. Rev. Lett.*, **66**, 1123–1126.

Smale, S. (1967) 'Differentiable dynamical systems', *Bull. Amer. Math. Soc.*, **73**, 747–774.

Smith, G.D. (1969) *Numerical Solution of Partial Differential Equations*, Oxford University Press, Oxford.

Sparrow, C. (1982) *The Lorenz Equation*, Springer, New York.

Steeb, W.-H. (1991) *A Handbook of Terms Used in Chaos and Quantum Chaos*, BI Wissenschaftsverlag, Mannheim.

Steeb, W.-H. and Euler, N. (1988) *Nonlinear evolution equations and Painleve test*, World Scientific, Singapore.

Stoker, J.J. (1959) *Nonlinear Vibrations*, Interscience Publisher, New York.

Strogatz, S.H. (1994) *Nonlinear dynamics and chaos: with applications to physics, biology, chemistry and engineering*, Addison-Wesley, Reading, Massachusetts.

Sussman, G.J. and Wisdom J. (1992) 'Chaotic evolution of the solar system', *Science*, **257**, 56–62.

Swinney, H.L. (1983) 'Observations of order and chaos in nonlinear systems', *Physica*, **7D**, 3–15.

Takens, F. (1981) 'Detecting strange attractors in turbulence', in: D.A. Rand and L.-S. Young (Eds.) *Dynamical systems and turbulence – Warwick 1980*, Lecture Notes in Mathematics, **898**, Springer, New York.

Tel, T. (1991) 'Controlling transient chaos', *J. Phys. A.*, **24**, 11359–11365.

Thom, R. (1975) *Structural Stability and Morphogenesis*, W.A. Benjamin, New York.

Thompson, J.M.T. (1992) 'Global unpredictability in nonlinear dynamics: capture, dispersal and the indeterminate bifurcations', *Physica*, **D 58**, 260–272.

Thompson, J.M.T., (1989) 'Chaotic phenomena triggering escape from a potential well', *Proc. Roy. Soc. London*, **A 421**, 195–225.

Thompson, J.M.T. and Bishop, S.R. (1994) *Nonlinearities and Chaos in Engineering Dynamics*, John Wiley & Sons, Chichester.

Thompson, J.M.T. and Stewart, H.B. (1986) *Nonlinear Dynamics and Chaos*, John Wiley & Sons, Chichester.

Thomsen, J.J. (1997) *Vibrations and Stability*, McGraw–Hill, London.

Ueda, Y. (1979) 'Randomly transitional phenomena in the system governed by Duffing's equation', *J. Stat. Phys.*, **20**, 181–196.

Ueda, Y. (1992) *The Road to Chaos*, Ariel Press, Santa Cruz.

Valis, G.K. (1988) 'Conceptual models of El Nino and the southern oscillation', *J. Geophys. Res.*, **93**, 13979–14019.

Van Campen, D.H. (1997) *Interaction Between Dynamics and Chaos in Advanced Mechanical Systems*, Kluwer, Amsterdam.

Walters, P. (1982) *An Introduction to Ergodic Theory*, Springer, Berlin.

Weickmann, K. M., (1991) 'El. Nino/Southern Oscillation and Madden Julian oscillations during 1981–82', *J. Geophys. Res.*, **96** Supp, 3187–3199.

Wiggins, S. (1988) Global Bifurcations and Chaos, Springer, New York.

Wiggins, S. (1990) *Introduction to Nonlinear Dynamics and Chaos*, Springer, New York.

Whitney, H. (1936) 'Differentiable manifolds', *Ann. Math.*, **37**, 645–708.

Winfree, A.T. (1988) *When Time Breaks Down*, Princeton University Press, Princeton.

Wolf, A., Swift, J., Swinney, H. and Vastano, A. (1985) 'Determining Lyapunov exponents from time series', *Physica*, **16D**, 285–314.

Wolfram, S. (1986) *Theory and Applications of Cellular Automata*, World Scientific, Singapore.

Zeeman, E.C. (1977) *Catastrophe Theory: Selected Papers 1972–1977*, Addison-Wesley, Reading, MA.